AUSTRALIAN SHIPWRECKS

A Pictorial History

My dear cousin Enricos,
I hope that you enjoy reading
this book as much as I enjoyed
writing it. Perhaps even read
some to the boys!
Best wishes,
Peter
2016

Peter Christopher

ISBN: 978 1 8647658 8 5

Unit 2, 1 Union Street, Stepney, South Australia, 5069.

Layout and design Anna Dimasi

Axiom Australia
www.axiompublishing.com.au

Layout and production in Australia

Printed in Malaysia

Contents

Display from Victoria's Twelve Apostles Visitor Centre on the Great Ocean Road. Part of the Historic Shipwreck Trail. Poem by Suzanne Howard

Introduction

Australia has a wealth of maritime history, having been at various times host to explorers, whalers, giant square-riggers and small coastal ketches, and in latter years, steamships. In addition the magnificent Murray River had fleets servicing inland waters.

Many of these vessels came from foreign ports, particularly those in the grain and passenger trade and at the time of the gold rushes.

The pre-colonial era saw our coast visited by explorers and traders, often without knowledge of their location.

Tragically, many met untimely ends, often with heavy loss of life. The era of sail, with all of its romance, was also an era of hardship with vessels being subject to the elements. Harsh sea conditions, when combined with limited navigational aids, virtually no communications and limited rescue facilities set the scene for disaster. In all, over nine thousand vessels have been lost on Australia's shores and in its river systems.

The exact number of vessels lost is uncertain, particularly when accurate records did not exist for many smaller local vessels. In other cases a ship travelling to or from Australia never reached its destination and the location of its loss is unknown. In some situations vessels were refloated, but cargo or anchors were left at the site, often ditched during refloating efforts. Many "wrecks" result from abandonment, not misadventure, and as a consequence some records do not show them as shipwrecks.

The book provides information about several of the more famous shipwrecks that have occurred in Australia.

A judgement was necessary as to which wrecks to include and which ones to omit. There is no right or wrong way of determining this. The approach that I have adopted is to aim for a book that is interesting to readers, with information and photos which give an insight to this important aspect of our maritime heritage, while not setting out to be an encyclopedia of shipwrecks.

It is not even possible to include all major wrecks. Rather, an attempt has been made to identify those that are both famous in their own right and also representative of certain circumstances, such as heavy loss of life, location, or those that carried cargo or emigrants.

The pre-colonial wrecks particularly have an importance which is absolutely unique. Their stories are an important part of the history of the discovery of Australia, but in addition they are fascinating accounts of tragedy, survival, hope, and in some cases, mystery.

The recognised maritime standard for ship measurements is traditionally Imperial, but as many readers are no longer familiar with this I have generally used metric. Imperial has been retained for tonnage. A ton, in maritime terms, is a measure of volume not weight, with one ton being 100 cubic feet. One foot equals 30.5 centimetres.

Names of both ships and people sometimes have more than one spelling, often when coming from a foreign port. Generally I have used the most common Anglicised version within this book, and make only a reference to alternative spellings. Referring to the early Dutch wreck on the Western Australian coast, for example, I refer to the *Gilt Dragon* rather than the *Vergulde Draeck*.

The story of shipwrecks is also a social history of the development of Australia. The early Dutch wrecks are testament of the expansion of world trade between Europe and Asia.

Wrecks of sailing ships also reflect the growth of Australian coastal towns, with the much later building of lighthouses at dangerous locations.

The loss of many women and children reflects the history of immigration as families journeyed on the ships together with sailors.

Reports of lives lost can often be inaccurate, in some cases due to births and deaths during a long journey confusing actual numbers aboard.

Much of my research has been from secondary sources and I gratefully acknowledge the work of others, particularly those listed in the bibliography at the rear of this book. I also strongly recommend to anyone wanting more information that they use the bibliography as a starting point.

Full listings of all wrecks are available both online and in published form, and I encourage readers to further explore this fascinating aspect of our maritime heritage.

Legal protection under the Historic Shipwrecks Acts of the States and the Commonwealth applies to wrecks and artefacts over seventy five years old as well as more recent wrecks satisfying required criteria.

While this book is about shipwrecks, it is also very much about social history. The circumstances that led the ships to be in our waters, and what happened to those ships, their passengers and cargo shaped our nation.

Peter Christopher.

Adelaide

Early Exploration & Navigation

Once it was accepted that the earth was round, the theory rapidly emerged that there must be a great southern continent to balance the land masses of the northern hemisphere. Hardly a sound basis for exploration, but nevertheless it did put Australia on the map before it was discovered.

The year 1606 was important with Captain Willem Janszoon aboard the Dutch vessel *Duyfken* first charting about 300 kilometres of the northern tip of Cape York Peninsula. The *Duyfken's* crew were the first known Europeans to set foot on Australia.

Later the same year Spanish Captain Louis Vaez De Torres sailed through what is now known as Torres Strait to the north of Australia with two vessels, the *San Pedro* and *Los Tres Reyes*.

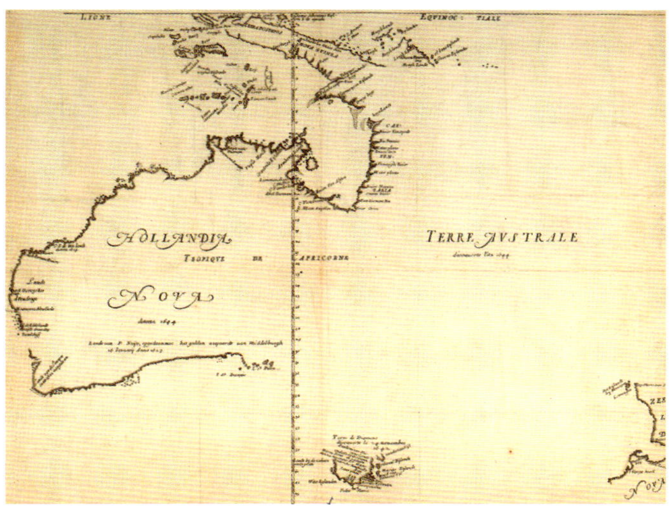

Controversy still exists in some quarters, with some firmly of the view that it was in fact Torres who first sighted Australia. It is generally accepted, however, that the tiny *Duyfken* and its crew were the first Europeans to do so.

In 1616 Captain Dirk Hartog in the Dutch *Eendracht* landed off the west coast, leaving behind a pewter plate marked with his name and date.

Over the next decade several Dutch ships sighted the coast of Western Australia and nearby islands, and sometimes land, but these are not voyages of exploration, but rather ships sailing too far west as they journeyed towards Batavia, as Jakarta was then known. Nevertheless, the sightings were recorded as hazards to assist future navigation.

The loss of the British ship *Tryall* off the Western Australian coast in 1622 with the loss of 93 lives becomes the first recorded European shipwreck. Forty Six survivors sailed north in two small boats.

An incident of note occurred in 1623 when the Dutch *Leijden* under the command of Captain Claes Hermanszoon made landfall south of Dirk Hartog Island and the first European baby was born on Australian soil.

Many Dutch ships continued to sight the western coast, with several lost there. The losses saw heavy loss of life and horrific circumstances, and are detailed within this book. In total the Dutch East India Company lost over 650 vessels around the world, yet its profits were such that these losses could be sustained. Several lost ships remain unaccounted for and could still be discovered on the Australian coastline.

Sailing from Batavia in 1642 the Dutch Captain Abel Tasman in the *Heemskerck* and *Zeehaen* discovered the west coast on Van Diemen's Land, now Tasmania. It was to be 1798 before George Bass and Matthew Flinders circumnavigated Van Diemen's Land in the small sailing boat *Norfolk*, finally establishing that it was an island, and not joined to the mainland.

The French vessels *Boudeuse* and *Etoile* under Captain Louis Antoine de Bougainville were stopped from reaching the east coast by the Great Barrier Reef in 1768, and sailed north.

It was 1770 when Captain James Cook in the *Endeavour* charted the east coast, including Botany Bay, and formally claimed his discovery in the name of England, renaming it New South Wales from its earlier name of New Holland.

English and French navigators continued to explore the new found land, and to chart its extensive coastline, with Captains Matthew Flinders, *Investigator*, and Nicholas Baudin , *Le Geographe* and *Le Naturaliste*, documenting the coastline between 1801 and 1803. Flinders circumnavigated the continent in the I*nvestigator*, and also proposed that the land be called Australia. His proposal for the name was adopted in 1824.

Le Geographe and *Le Naturaliste* 1802 in Timor

Even as details of the coast were recorded hundreds of ships were lost because of poor navigation equipment and lack of maritime aids such as lighthouses. Combined with treacherous reefs and the forces of nature it is testament to the skills of their crews that so many ships survived.

Added to all this, elements remain of the unknown. The tale of the Mahogany Ship off the coast of Victoria near Warrnambool raises significant possibilities that all of the above recorded discoveries may in fact have been preceded by other unknown explorers.

1663 veiw of the world by Nicolaes Visscher showing Australia's South West.

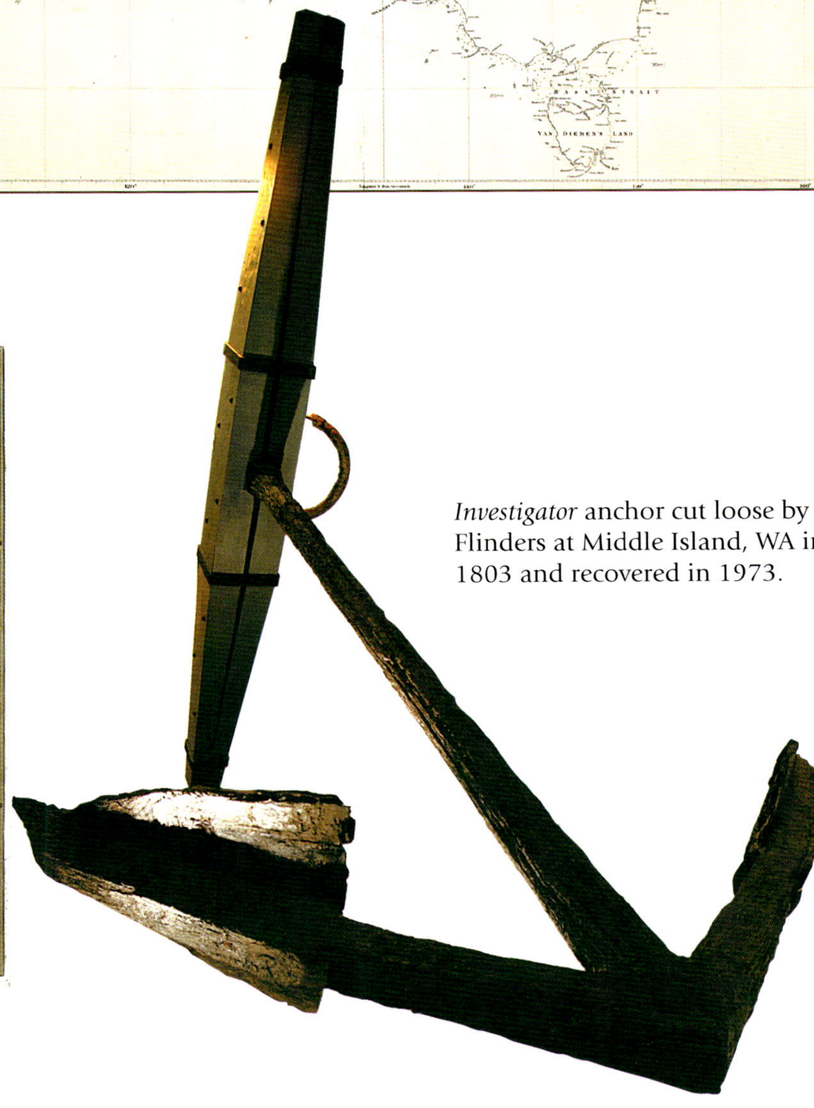

Investigator anchor cut loose by Flinders at Middle Island, WA in 1803 and recovered in 1973.

Æquator
t'Amsterdam
G.F.de Wits landt
Willems Revier
't Land vand'Eendracht
Dirck Hartogs Ree
Houtmans Abrolhos
I.d'Edels landt
't Land van de Leeuwin
't Landt van P. Nuyts
opgedaen met 't Gulden Zeepaerdt van Middelburgh
26 Ianuary A°. 1627
I. S. Francois
I. S. Pieter
Tropicus Capricorni

WESTERN AUSTRALIA THE EARLIEST WRECKS

Batavia

The loss of the Dutch East India Company ship *Batavia* off the Western Australian coast in 1629 remains the most shocking of all wrecks in Australian waters.

Not only were the 40 drownings when *Batavia* sank tragic, but the mutiny and murders that followed remain unparalleled in our maritime history. 125 people were killed by mutineers in the most appalling and shocking circumstances following the loss of the *Batavia*.

The *Batavia*, under the command of Captain Jacobsz, left Holland with 2 other ships for Java, what we now know as Indonesia, on 26 March 1629. Among the 316 people aboard the *Batavia* was Captain Francisco Pelsaert, the head of the fleet.

The Dutch East India Company, or Vereenighde Oost-Indische Compagnie (VOC), had been established in 1602 and was at the time the world's largest trading company, with a virtual monopoly on the lucrative spice trade with Asia. In 1629 it sent eight other ships to the East Indies, as well as the *Batavia* fleet.

The *Batavia* was separated from the other vessels during a storm which struck after they left the Cape of Good Hope, the southern tip of Africa. Now alone, *Batavia* continued towards its original destination of Java, but had drifted off course.

On 4 June *Batavia* ran onto a reef 50 kilometres off the Western Australian coast. It was clear by the following morning that the ship would soon break up and boats started ferrying those aboard to a nearby island, now known as Beacon Island, in the Abrolhos Islands group..

Before the evacuation was completed the seas became so rough that further rescue was impossible. *Batavia* broke up a week later and sank. Those still aboard drowned.

Limited food, water and other supplies had been saved. As there was only an inadequate supply of natural water on the island and no source of food other than what had been saved from the ship, Pelsaert determined to search the other nearby islands and the mainland.

Together with Captain Jacobsz and over 40 others in two boats, Pelsaert sailed on this search but it was to be two weeks and hundreds of kilometres before water was found. Because of their distance from the *Batavia* wreck site they decided to sail on to Java to arrange for a rescue boat rather than return to the stranded passengers and crew. While setting out on a mission for water and a rescue boat, they are regarded by many as abandoning the survivors.

Arriving in Java Pelsaert arranged for the frigate *Sardam*, part of his fleet from Holland, to leave within a week on a rescue mission, and to try and recover as much of the *Batavia's* cargo as possible. The cargo included 12 chests of silver coins, as well as jewels and other trade goods.

Jacobsz was arrested in Java as the captain responsible for the loss of the *Batavia*.

During the time Pelsaert and Jacobsz were away, a crew member, Jeronimus Cornelisz, initiated a plan to take over the rescue ship he expected to arrive for the survivors. The mutiny had been planned before the sinking, with Cornelisz originally plotting to seize the *Batavia* and its valuable cargo before it reached Java.

Cornelisz, a non swimmer, had remained on the broken up Batavia for 8 days, when he finally was washed onto the shore clinging to a spar after the last remnants of the *Batavia* disappeared.

As *Batavia* broke up water casks, food, timbers and sails also washed up and were used by the castaways. As well as erecting tents from sails, timber was used to construct boats capable of reaching neighbouring islands.

Food such as birds and seals and water were available on nearby islands, and the survivors were now able to access them.

While Cornelisz planned to take over the rescue boat with his mutineers he knew that there would be strong resistance from soldiers and many passengers. Cornelisz was the senior VOC officer amongst the castaways and used this authority to send the soldiers under the command of Weibbe Hayes off searching for water on another island, while separating the remaining crew and passengers between other islands.

Beacon Island, also known as Batavia's Graveyard, was the mutineers' base and where the killings began, with the strongest killed first and at night. Their absence was explained by the story that they had gone to other islands.

The mutineers then started killing the smaller groups on each of the islands. Some, however, escaped and using timbers managed to get to the island that Weibbe Hayes and his soldiers had been left on. Unexpectedly water had been found and this made possible their survival.

Meanwhile the killings continued with beheading, drowning, stabbing and bludgeoning of skulls of men, women and children common means of death, although a number of women were spared for the amusement of the mutineers.

Twenty two of Cornelisz men were sent to kill the unarmed soldiers, but did not succeed. Cornelisz himself then attempted to do so with more men, but was himself captured. Weibbe Hayes had organised a defence and made makeshift weapons for his group of forty seven using sharpened iron from barrel hoops and clubs from timber that had washed up from the wreck.

Hayes took to a boat when the *Sardam* came into sight soon after, and warned Pelsaert of the mutiny. When the mutineers' boats approached the *Sardam* Cornelisz' men were captured. Seven of the mutineers were tried and executed on the islands, and two were marooned on the mainland. The remainder were taken to Jakarta.

Divers from the *Sardam* recovered eleven of the 12 chests of silver coins. The chest of jewels had been taken ashore by Pelsaert following the *Batavia* running onto Morning Reef.

Batavia was thought for many years to be in the southern Abrolhos group of islands, and it was only when Henrietta Drake-Brockman arranged for the translation of Pelsaert's journal that the search became focussed on the more northern Wallabi group of islands. The location of the wreck was rediscovered in 1963, and was subsequently the site of a major maritime archaeological project by the West Australian Maritime Museum.

In addition to recording vital information, a significant amount of artefacts were recovered, including cannons, coins and pottery. As well, part of *Batavia's* hull was raised in cut up sections, and following treatment to prevent deterioration, was re-assembled and is now on display at the Museum. 137 large pre-cut stones, whose purpose remained a mystery, were also raised from the wrecksite. Eventually after the location of a drawing it was determined that these were a portico ready for assembly over the entrance to a fort in the city of Batavia, now Jakarta.

The *Batavia* remains one of Australia's most significant shipwrecks.

Beacon Island camp of *Batavia* survivors.

WAMM survey of *Batavia* wreck site.

Re-assembled portico destined for fort entrance in the city of Batavia (now Jakarta).

Batavia hull timbers.

Batavia victim. Cutting wound evident on top of skull.

Batavia cannon. Longest of five bronze guns recovered, displaying the coat of arms of the Admiralty of Rotterdam.

Pelsaert Silver

"On Pelsaert's advice, the *Batavia* was carrying some 'experimental' trade items. This finely engraved ewer was produced at Pelsaert's request to provide luxury items for the Indian market. It was probably destined for an Islamic hand-washing ceremony. In one of the cartouches such a ceremony is depicted." WAMM

Tryall

The earliest recorded wreck on the Australian coast was the British ship *Tryall*, lost off the Western Australian coast on 24 May 1622, whilst en route to Batavia, having left Plymouth in September 1621. Some reports show the ship's name as *Tryal* or *Trial*.

As the spice trade demanded speedy trips to enable quick profits, every effort was to be made to travel by the fastest route, and Brookes was required to travel in the southern latitudes with the objective of cutting several months off the journey.

Stopping at Cape Town to resupply, Captain Brookes also took on Thomas Bright as first mate as he had undertaken the route previously. *Tryall* left Cape Town on 19 March 1622.

Tryall was following the Brouwer route to the spice islands around Java, which involved sailing east from the Cape of Good Hope along a low latitude to get the benefit of the Roaring Forties, then sailing north.

To purchase the valuable spice, *Tryall* carried a significant amount of silver coinage.

Unfortunately Captain Brookes sailed too far east before changing direction. Although the Australian coastline had been sighted from 1 May, northerly winds prevented the *Tryall* from sailing north to Java. Brookes tragically remained in the vicinity of the coast for far too long.

Lost after striking a reef in the region of the Montebello Islands, *Tryall* quickly sank and 93 crew perished. Captain James Brookes survived, together with 45 others in two boats.

The survivors journey of almost 2,000 kilometres was difficult, with 10 persons in a small skiff and 36 in a longboat, with little water and limited supplies. One person in the skiff died.

After reaching Java on 5 July, Captain Brookes gave an account aimed at removing claims of negligence against him by describing the reef as being much further out to sea than it actually was, and wrongly blaming the mapping of earlier navigators.

The truth became known with the account of Thomas Bright, who had taken charge of the second boat of survivors. Bright believed Brookes to be responsible for the loss of the *Tryall* and many of its crew, and expressed his version of events in a secret letter to an English East India Company representative.

Bright also accused Brookes of leaving the wrecksite before attempting a proper rescue of other crew. Brookes version of events was that the *Tryall* sank within 3 hours, whereas Bright's report was that it did not fully sink for over 10 hours, and that Brookes' skiff could have taken on more survivors.

The wreck was discovered in 1969, with cannons and anchors visible on the site. Bright's letter, written over three hundred years earlier, was instrumental in the wreck's discovery.

A Cannon from the *Tryall*

This cast iron cannon was recovered in 1985 from the English East India Company ship *Tryall*, the earliest known shipwreck in Australian waters. While on a voyage from Plymouth to Batavia (Jakarta) in 1622, *Tryall* struck an uncharted reef north of the Montebello Islands.

Ninety-three sailors were abandoned when the master John Brookes and forty-five others sailed away in two boats, eventually arriving at Batavia. En route, one of these boats landed on the Montebello Islands in search of water, making it the first known English landing on Australian soil.

The broad arrow symbol established the cannon as British.

Zeewijk

Almost a century after the loss of the *Batavia*, the Dutch East India Company's vessel *Zeewijk* was also lost on the reefs of the Houtman's Abrolhos Islands.

Built in 1725, the 48 metre *Zeewijk* left Amsterdam on 7 November 1726 for Batavia with a crew of 212 under the command of Captain Jan Stevns. Ten chests of guilders made up part of the cargo, and she had 36 cannon.

After resupply at the Cape of Good Hope, *Zeewijk* set sail again on 21 April. Six weeks later, at 7.00pm on the evening of 9 June *Zeewijk* struck a reef. Breakers seen by the lookout 30 minutes earlier had been thought to be reflections of the moon, and ignored.

The *Zeewijk* did not break up immediately, but conditions prevented the launch of the long boat for a week. Eventually the boat was launched and the crew left the wreck and moved to a nearby island, later named Gun Island.

Once the crew and salvaged cargo were on the island, the long boat was prepared to sail to Batavia to arrange rescue, with First Officer Peter Langeweg in command. They never reached their destination.

Using timbers from the wrecked *Zeewijk*, the remaining crew constructed another vessel, named *Sloepie*, and sailed for Batavia in March 1728. On 21 April, one year after their departure from the Cape of Good Hope, the surviving 82 crew reached Sunda Strait. Captain Stevns was prosecuted as the officer responsible for the loss of the *Zeewijk*.

In all 130 crew had perished.

In 1840 relics of the wreck were found by the HMS *Beagle*, including a brass swivel cannon found on the island by surveyor, John Stokes. Fifty years later more relics, including bottles and cooking pots, were located on the island by guano miners.

The subsequent discovery of cannons and anchors occurred after a Navy Officer received reports in 1952 that a cray-fisherman had discovered cannons on the western side of the Abrolhos Islands. It was to be 1968 before Western Australian diver and, now author, Hugh Edwards took Museum staff to the main site of the wreck. Extensive surveys have been undertaken since.

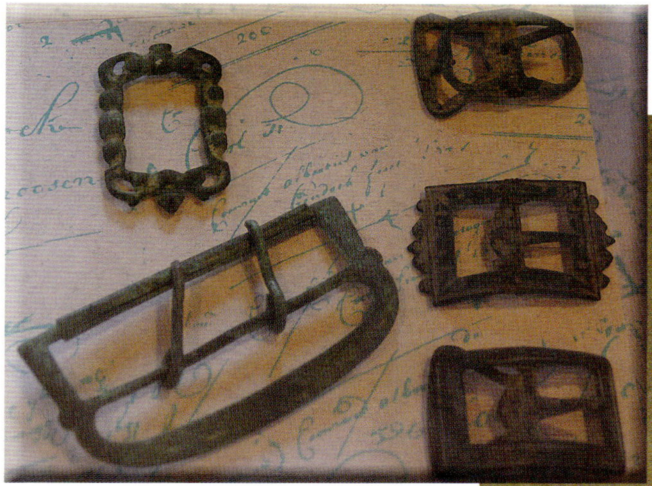

Zeewijk bottles, and baldric buckles used to carry a sword or cutlass.

Gilt Dragon

The discovery of the location of the shipwrecked *Gilt Dragon* in 1963 achieved what had not been possible over 300 years earlier, the actual finding of the wrecksite.

Lost in 1656 after striking a reef on 28 April, seven of the *Gilt Dragon* crew were sent in a small boat to try and reach Batavia over 2,200 kilometres to the north. They reached their destination 40 days after the loss of their ship, but rescue attempts failed to find the remaining survivors.

The 42 metre *Gilt Dragon* was a Dutch East India Company vessel on its second voyage to Batavia when it struck a reef only 5 kilometres off the Western Australian coast, about 100 kilometres north of Perth.

118 of the crew died in the wreck, with 75 survivors, including the vessel's Captain, Pieter Albertsz, making it to the mainland in 2 small boats. One of these boats made the subsequent journey north seeking rescue for the remaining 68 crew.

The *Gilt Dragon*, or *Vergulde Draeck*, had been carrying 8 chests of silver coins and other valuable cargo to be used for the purchase of the much valued spices sought after by Europe.

Two rescue boats, the *Goede Hoop* and the *Witte Valke* sailed south to search for survivors, and the *Gilt Dragon's* valuable cargo, but found no trace of either. Eleven of the *Goede Hoop's* crew were lost when their boat was overturned as they attempted to reach shore on the Australian coast.

Unsuccessful searches occurred over subsequent years, with further loss of life in 1n 1858 as a boat from the *Waeckende Boey* was separated from its ship. Only 4 of the 14 on board the boat survived after a journey north to Batavia lasting 6 months.

In 1931 a boy playing in the sandhills near the mouth of the Moore River discovered about 40 Dutch coins dated between 1619 and 1655 leading historian Henrietta Drake-Brockman to speculate that the wreck of the *Gilt Dragon* was in the vicinity.

The *Gilt Dragon* was discovered by spearfishermen in 1963, the first of the pre-colonial Dutch wrecks to be found. The *Batavia* was found later the same year.

Legislation to protect the site was enacted in 1964 after looting took place, including significant damage through the use of explosives. These treasure seeking vandals clearly had no interest in the ship remains or other fragile historic relics.

The Western Australian Maritime Museum became the centre of maritime archaeology in Australia as a result of these and other discoveries, and now has a well deserved international reputation in this field.

The *Gilt Dragon* site has yielded a range of 17th century artefacts, including navigation equipment, clay pipes, jugs, ceramic masks, brass utensils, tools, and bottles, as well as silver coins and elephant tusks.

The jugs, also found on the *Batavia*, included what are now often referred to as "beardmen jugs" or "Bellarmine jugs", carrying a bearded man's face on their neck. One version of the history of these jugs is that Cardinal Bellarmine (1542 -1621) was ridiculed by the Protestants opposed to his sermons by a caricature of his face being placed on beer jugs. Drinkers would mockingly toast Cardinal Bellarmine as they took a swig from the jug bearing his image.

Bellarmine jug and mortar and pestle.

Elephant tusk from *Gilt Dragon*.

Zuytdorp

It is possible that some of the *Zuytdorp's* crew survived the loss of the Dutch East India Company vessel off the remote Western Australia coast in 1712, and became the first European settlers.

Tom Pepper, a stockman in the region south of Shark Bay, found silver coins at an old campsite on top of high cliffs in 1927. In 1954 he told a geologist working in the area, Phillip Playford, of his find, and showed him the site and artefacts.

The coins were established as Dutch, and a date on one of them showed it was minted in 1711. Records in the Netherlands showed that only two ships carried these particular coins from that minting, with one of them having been lost without trace. That ship was the *Zuytdorp*.

Following the determination of the source of the coins, local divers searched the waters below the cliffs where Tom Pepper had found the original silver coins many years before. They soon located ships timbers, cannon and anchors, as well as thousands of silver coins. The description "carpet of coins" quickly came to be the common description of this find in the media.

The site is on a rugged and remote part of the coast, about 40 kilometres north of the Murchison River.

Fear of looters led the WA Maritime Museum to set up a watch over the site, with a caravan at the top of the cliffs. During a short absence by the watchman the museum caravan was firebombed, and because of the risk posed, the Museum stopped its diving and guarding activity for five years. The treasure was looted during this time.

An ongoing mystery is the fate of the crew of the *Zuytdorp*. The camp and coins above the cliffs make it clear that there were survivors, with theories also that they may have been helped by the local Aboriginal people.

Significant and heavy fittings from the ship could only have been recovered by the ship remaining accessible above the reef that it struck, and there being sufficient numbers of survivors to move items such as 9 heavy breech blocks associated with the ship's swivel cannons that were located ashore.

An American woman contacted Phillip Playford in 1988, telling him that her Aboriginal husband had died earlier. The cause of his death was a believed to be a genetic disease, variegate porphyria, linked to South African Afrikanners. Over a hundred of the *Zuytdorp's* crew were taken on in Cape Town, and could have carried the disease.

DNA analysis in both Australia and South Africa, however, concluded that shipwrecked sailors were not the cause of the disease identified amongst the Australian Aboriginal population.

NEW SOUTH WALES

As the location of Australia's first planned European settlement Sydney was the destination for numerous passenger and cargo ships, and consequently shipwrecks.

These vessels from overseas were large, capable of undertaking journey's of many months from Europe initially, and subsequently Asia and America.

As the colony grew the population spread along the coast, with shipping being the most common form of transport for movement of cargo. Roads were of poor standard or non existent, and horse or oxen wagons had only limited carrying capacity.

Ships could move people and vast quantities of cargo quickly, but the combination of a rugged coastline and lack of maritime aids resulted in the loss of many of these vessels, both large and small.

Melbourne was the main destination port in Victoria, and the majority of significant wrecks in that State were of large sailing ships from overseas sailing to Port Phillip Bay. Even the large sailing ships bringing passengers and cargo from overseas for Sydney most often stopped at Melbourne en route or sailed along the Victorian coast and through Bass Strait before turning north towards Port Jackson, encountering the dangers of that region. Many came to grief.

Unlike Victoria which did not have as extensive a coastal population and trade, New South Wales developed major coastal centres.

As the population spread and coastal centres grew in size, smaller vessels started being used as well as their giant ocean crossing counterparts. The development of steam power saw a transition of coastal shipping from sail to steam, and this was reflected in the type of wrecks occurring. These were initially a mix of Paddle Steamers and propeller driven vessels, although paddles did not prove to be as effective in the ocean with its waves and swells as they were on calmer rivers.

The New South Wales coast has thousands of wrecks, with over three hundred in the coal mining and ship building region of Newcastle alone.

Sole survivor, James Johnson

Dunbar

Sydney's most famous wreck is without doubt the wooden clipper ship *Dunbar*, lost in 1857 just off South Head, at the entrance to Sydney Harbour.

Shipowner Duncan Dunbar placed an order for a ship capable of carrying passengers, as well as a large amount of cargo, with James Laing and Sons, to meet the demands generated by the Australian goldrushes.

Taking 16 months to build, the 1167 ton *Dunbar*, was completed in 1854, the largest vessel built in the shipyards of Sunderland to that time.

On its second trip to Australia, and with a crew of 59 and 63 passengers, *Dunbar* made good time arriving off the New South Wales coast in eighty one days. *Dunbar's* master, Captain Green, had sailed the trip seven times before.

The cargo included furniture, machinery, food, alcohol and cutlery, as well as ink dye for the colony's first postage stamps.

After good weather and a swift trip, having travelled through Bass Strait and then along the east coast, the situation changed as the ship approached Sydney from the south. Heavy rain fell and the seas built as darkness descended, but despite the conditions Captain Green believed that he knew the area well enough to sail into the protection offered by the harbour on the night of 20 August.

The decision not to wait until the light of morning doomed the ship.

Lookouts were posted to watch for North Head. Once sighted, the Captain would know that they had cleared South Head, and that the ship could change to a westerly heading and enter the harbour.

A cry of "breakers ahead" by the second mate resulted in the ship turning west, but the breakers were not those of North Head as anticipated, but off South Head. The result was that instead of turning safely into the harbour, the *Dunbar* went into the cliffs south of the entrance.

The ship quickly began to break up, topmasts were brought down by the impact, and lifeboats could not be launched, with the ship at an acute angle laying broadside under the cliffs. Passengers and crew alike were thrown into the water and drowned as they and their ship were pounded into the cliffs.

As the ship broke up and masts and rigging filled the water, those that did not immediately drown were killed by the debris or smashed upon the jagged cliffs, as evidenced by the mutilated bodies subsequently recovered.

James Johnson, a young crew member, managed to stay afloat on some boards together with another sailor. Although Johnson was able to climb onto a rocky ledge, the second sailor was washed back into the sea.

On 22 August, the second morning after the *Dunbar* was lost, Johnson's cries for help were heard by searchers who had come to the area as news of the disaster swept through Sydney. He was the only survivor, with 121 perishing.

Bodies were recovered from the water and buried at St Stephen's Cemetery, Newtown. The magnitude of the disaster and its impact can be judged by the fact that 20,000 people lined Sydney's George Street for the funeral procession on 24 August. Many of these people had been expecting to be celebrating the arrival of friends, family, loved ones and associates. Many had instead faced the grizzly task of identifying them.

Today a plaque and anchor above the cliffs mark the location of the *Dunbar*, as well as the memorial within the cemetery. A collection of items from the wreck are displayed at the Australian National Maritime Museum.

After the barque *Catherine Adamson* was wrecked on North Head with the loss of twenty one lives two months later the decision was made to construct the Hornby Lighthouse on South Head.

In 1866, James Johnson was instrumental in rescuing the sole survivor of the Paddle Steamer *Cawarra* off Newcastle, where he was a lighthouse keeper.

Dunbar anchor.

The shore is literally white with candles, and the rocks covered or so deep with articles of every kind – boots, panama hats and bonnets are here in abundance. Drums of figs, hams, pork, raisins, drapery, boots and pieces of timber are piled in heaps along with the keel of the Dunbar,' said the Sydney Morning Herald on 23 August 1857.

'The rumours as to the fact of a dreadful shipwreck having just occurred soon assumed distinct shape and certainty. At length it generally became known in Sydney that numerous dead and mutilated bodies of men, women and children were to be seen floating in the heavy surf at the Gap thrown by immense waves at a great height; and dashed pitilessly against the rugged cliffs, the returning water sweeping them from the agonised sight of the horrified spectators …'
(*A narrative of the melancholy wreck of the Dunbar,* James Fryer, Sydney 1857).

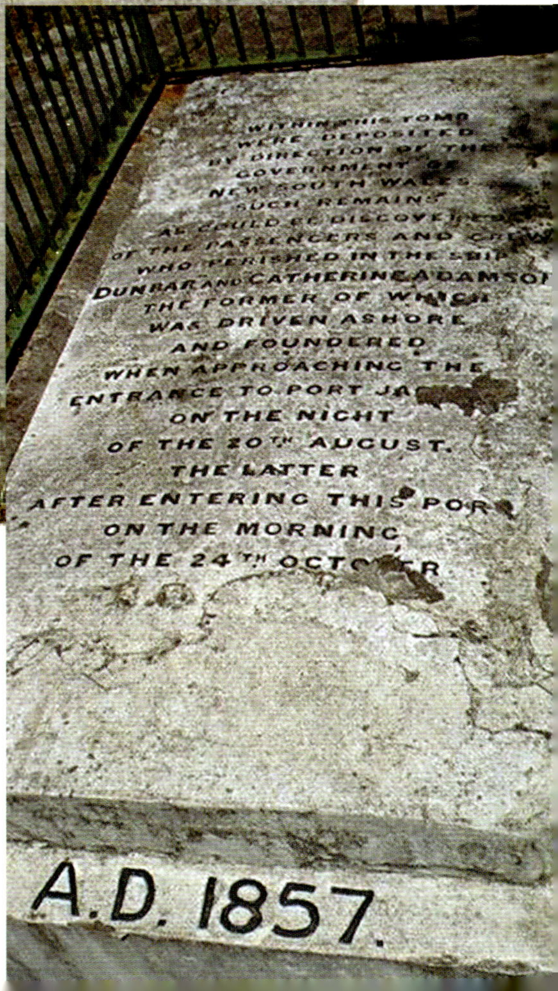

Tombstones for victims of both *Dunbar* and *Catherine Adamson* St Stephen's Cemetery.

Maitland

Built in 1870 in Glasgow, the 70 metre iron Paddle Steamer *Maitland* regularly traded along the New South Wales coast, before rail joined Sydney and Newcastle.

Leaving Sydney at 11.00pm on 5 May 1898 with 63 crew and passengers, *Maitland* ran into severe weather even before it got out of the harbour and through the Heads.

Maitland's cargo included sugar, rice, iron, tea and alcohol.

The deckhouse was carried away by the huge waves created by the gale, and the hull started rapidly taking water. In an effort to stay afloat and not capsize Captain Skinner ordered cargo to be thrown overboard.

The huge seas flooded the boat and with the engine fires extinguished by the water, *Maitland* was helplessly adrift and at the mercy of the elements as it drifted through the night.

Maitland struck East Reef off Cape Three Points about 50 metres from shore shortly before 6.00am, breaking in two. The forward section quickly sank, taking many of the crew and passengers with it. Some made it to shore, suffering severe injuries from the reef, rocks and wreckage.

The survivors left in the stern section attempted to get ashore by running a line between the ship's remains and the shore. Three initial survivors were lost when the line broke.

A one year old baby on board, Daisy Hammond, was nursed through the night by the crew after her injured mother had been gotten ashore. Baby Daisy was carried to safety on the back of a crew member, boatswain Johanson, who made the short but treacherous escape from the stern section along the line which had been secured.

Daisy Stevens, as she became, lived to the age of 90 years, and visited the wreck site in 1971. After her death her ashes were returned from Canada where she had lived for some time and scattered at the wreck site.

Twenty Four passengers and crew died. Within weeks of its loss very little remained of the *Maitland*, which was quickly broken up by the seas on the reef upon which it had been driven. Awards for bravery were presented to many involved in the rescues by the Royal Shipwreck and Humane Society, including Mrs Anita Hammond, Daisy's mother.

The gale that sunk the *Maitland* was the fiercest for some decades, resulting in the loss of eleven other vessels. The area of the loss was renamed Maitland Bay.

f the Maitland

NATIONAL·SHIPWRECK·RELIEF·SOCIETY OF NEW SOUTH WALES.

This CERTIFICATE has been awarded to

WILLIAM WILLIAMS (AB. - SS "Maitland")

together with the Society's Silver Medal for courageously endeavouring to take a line from the wreck of the "Maitland" to the shore near Cape Three Points, Broken Bay, N.S.W. on the 6th May, 1898.

Granted 4th July, 1898

Chairman

Secretary

Top left Daisy Hammond in 1908 with her rescuer, boatswain Johanson, ten years after the *Maitland's* loss.

Top right Anita Hammond in later years.

Malabar

The loss of the Motor Vessel *Malabar* in 1931 led to the renaming of a Sydney suburb to carry its name two years later.

Although previously known as Brand, the suburb was generally referred to as Long Bay because of the prison of that name nearby, a fact that caused residents discomfort, and perhaps affected property values.

Malabar, a 4512 ton diesel powered cargo steamer, had left Melbourne on 31 March for its regular trip to Singapore and was travelling northwards along the NSW coast in heavy fog on 2 April 1931 when it ran onto the rugged Long Bay headland 12 kilometres south of Sydney.

Captain George Leslie had ordered the ship further out to sea because of the lack of visibility, but a misunderstanding by the helmsman resulted in the ship being turned 5 degrees to port, towards the coast, not away from it as intended.

The ship hit the rocks, with the bow high and dry, and all 109 crew and 28 passengers were able to safely get off the vessel in lifeboats. The sea was calm on the day, although blanketed in fog.

The ship's cat would not leave the vessel and was lost.

Captain Leslie tried to back the ship off, and an attempt by the *Charlie Cam* to tow it off was unsuccessful with the tow line

snapping. *Malabar* could not be refloated and within two days was broken up by the seas.

A Court of Marine Inquiry was convened and found Captain Leslie guilty of negligent navigation, cancelling his Certificate.

Half a million people are reported to have visited the site of the wreck in the week following its stranding, with the time including the Easter period.

Malabar itself was named after a town in Java, the most populous island in Indonesia and the location of its capital city, Jakarta. Jakarta, or Batavia as it was originally known, was the centre of the East Indies spice trade which saw many Dutch vessels visit it, with some lost on the Western Australian coast as they sailed towards it.

Cawarra

The *Paddle Steamer Cawarra* sailed from Sydney to Queensland with 63 passengers and crew, under command of Captain Chatfield in July 1866. The severity of a storm on the first night out was such that the Captain determined to seek safety of port and turned for Newcastle.

In an environment where the commercial viabilities of towns and industries relied on regular and fast carriage of people and cargo, most ships departed at their appointed time, in all but the most extreme of weather conditions.

Attempting to enter the channel *Cawarra* was forced by waves towards a reef known as the Oyster Bank, compelling the Captain to turn towards the relative safety of the open sea. Soon after waves over the bow swamped the engine room and *Cawarra* lost power and steerage, and she started to sink.

The side paddle wheels were vulnerable to easy damage in heavy seas, with the boxes that covered them able to be broken by massive waves and forced into the wheels themselves. In practical terms paddle steamers were far more suited to rivers and protected waters, not the ocean.

A lifeboat was launched but soon swamped drowning all but one crew member, Frederick Hedges, who managed to get back to the *Cawarra* joining a few remaining crew. Rescue from the shore was attempted but the boat could not get close to the *Cawarra* because of the size of the seas.

As the *Cawarra* finally sank fully, Hedges managed to stay afloat clutching to a piece of timber. He was carried into the channel where he got hold of a buoy, from which he was rescued by a boat from shore. One of the three people in the rescue boat was James Johnson, sole survivor of *Dunbar* wreck. Hedges was the only survivor of the *Cawarra*.

Cawarra was an iron paddle steamer of 552 tons, built in Glasgow in 1864 and arriving in Australia the same year.

Wreck of the *Cawarra*

Loch Ard Gorge, one of the most visited locations on the Great Ocean Road Historic Shipwreck Trail.

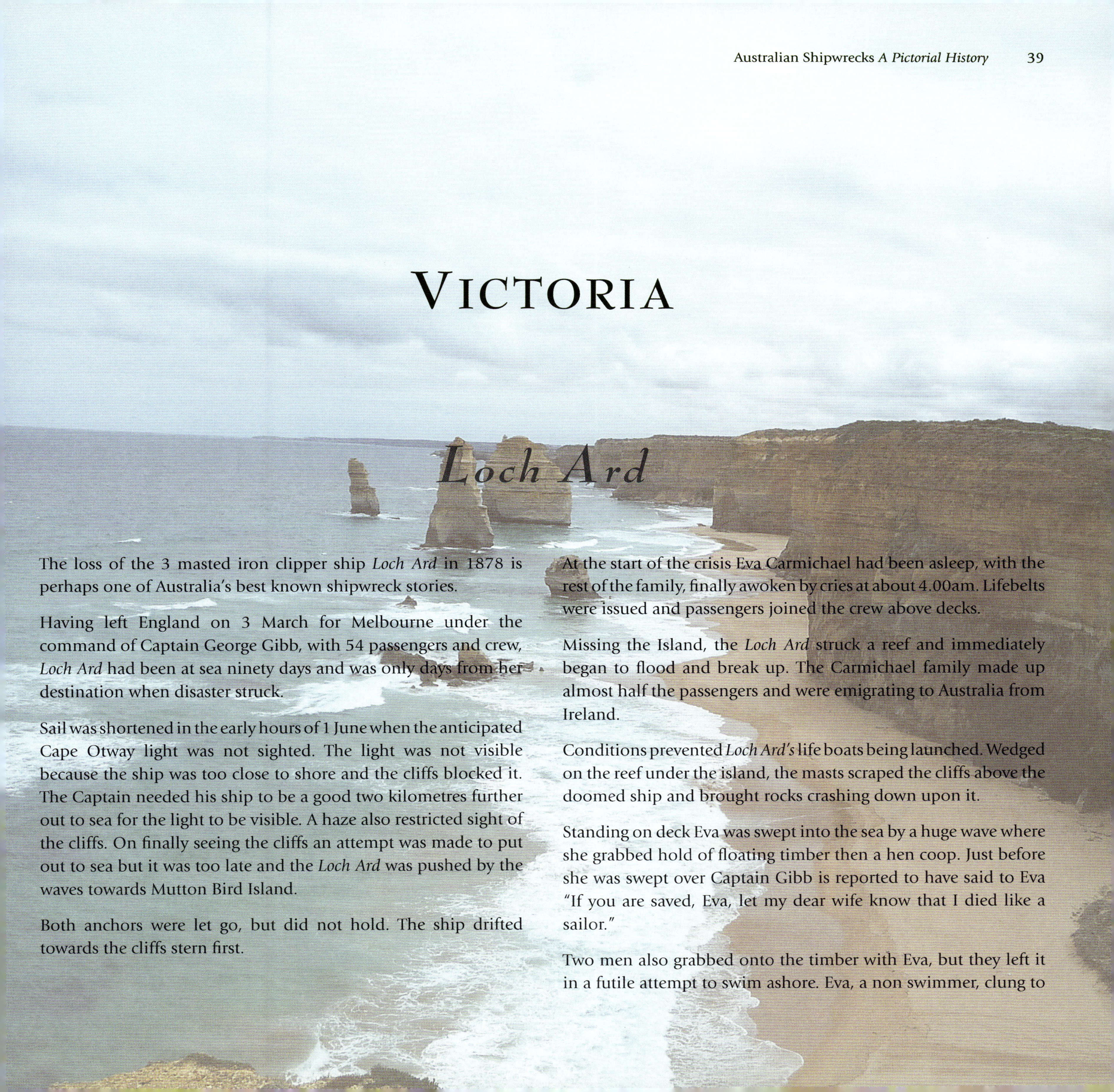

Victoria

Loch Ard

The loss of the 3 masted iron clipper ship *Loch Ard* in 1878 is perhaps one of Australia's best known shipwreck stories.

Having left England on 3 March for Melbourne under the command of Captain George Gibb, with 54 passengers and crew, *Loch Ard* had been at sea ninety days and was only days from her destination when disaster struck.

Sail was shortened in the early hours of 1 June when the anticipated Cape Otway light was not sighted. The light was not visible because the ship was too close to shore and the cliffs blocked it. The Captain needed his ship to be a good two kilometres further out to sea for the light to be visible. A haze also restricted sight of the cliffs. On finally seeing the cliffs an attempt was made to put out to sea but it was too late and the *Loch Ard* was pushed by the waves towards Mutton Bird Island.

Both anchors were let go, but did not hold. The ship drifted towards the cliffs stern first.

At the start of the crisis Eva Carmichael had been asleep, with the rest of the family, finally awoken by cries at about 4.00am. Lifebelts were issued and passengers joined the crew above decks.

Missing the Island, the *Loch Ard* struck a reef and immediately began to flood and break up. The Carmichael family made up almost half the passengers and were emigrating to Australia from Ireland.

Conditions prevented *Loch Ard's* life boats being launched. Wedged on the reef under the island, the masts scraped the cliffs above the doomed ship and brought rocks crashing down upon it.

Standing on deck Eva was swept into the sea by a huge wave where she grabbed hold of floating timber then a hen coop. Just before she was swept over Captain Gibb is reported to have said to Eva "If you are saved, Eva, let my dear wife know that I died like a sailor."

Two men also grabbed onto the timber with Eva, but they left it in a futile attempt to swim ashore. Eva, a non swimmer, clung to the timber and was washed towards shore in what is now named Loch Ard Gorge.

On the *Loch Ard* a lifeboat was eventually got into the water by 18 year old apprentice Tom Pearce and four other crew members. They had no sooner clambered into it when it was overturned, with Tom trapped under it.

Tom's account states "*I was under the boat and must have remained there about three quarters of an hour. I got from under, but could see nobody, but a lot of floating wreckage.*"

Eventually the upturned boat was washed towards shore, and into a narrow gorge between the cliffs. He righted the boat, but hit the cliffs and he was thrown out. By this time it was approaching daylight, and seeing a small beach, Tom swum towards it.

Reaching the shore, exhausted and with a gash to his head, he found refuge in a cave.

Later, while looking among the washed up wreckage for food, Tom heard a cry for help. He initially thought it was from above the cliffs, but after a second cry saw a struggling girl in the water. Eva had been washed towards shore but could not get to the beach. Swimming out through the floating wreckage to her, he managed to get hold of a nearby floating table, and after an hour, succeeded in reaching shore.

Eva later reported "*He took me into a wild looking cave a few hundred feet from the beach and finding a case of brandy, broke a bottle and made me swallow some, which revived me. He pulled some long grass and shrubs for me to lie on. I soon sank into a state of unconsciousness and must have remained so for hours. We must have been fully five hours in the water.*"

They were to be the only survivors.

Waking after a long sleep, and struggling to find a way up the surrounding cliffs, Tom finally made the top, but nothing was visible in any direction, and just kept walking. Mustering sheep from Glenample Station homestead, George Ford and 14 year old William Till came across Tom walking along the cliff top.

Ford rode for help and Tom returned to the wrecksite below the cliffs.

Glenample's part owner, Hugh Gibson, was unsure of the truth of the story, particularly with Tom reported as smelling of brandy, but decided to investigate. Eva was not in the cave, but later that night was found helpless in nearby scrub. One of her rescuers was William Shields, whose uncle it emerged was a pharmacist who had worked with her father, Doctor Evory Carmichael, in Dublin.

After a struggle to get Eva to the top of the cliffs the rescue party got back to Glenample sometime about 2.00am.

Two hours after finding Eva, George Ford left on a 72 kilometre ride to Camperdown to alert the necessary authorities, including the Commissioner of Customs, the Police and agents for the *Loch Ard's* owners.

Wreckage washed up onto the beach over coming days, forming a barrier over 2 metres high. It did not take long for other rescuers, authorities, reporters, curious onlookers and plunderers to arrive, with four hundred persons at the scene within days.

The fact that the only two survivors were a male and female in their late teens resulted in constant media speculation of romance. It was not to be, with Eva returning to Ireland where she subsequently married. Tom Pearce continued with a life at sea, becoming a Captain. Tragically his own son was lost on the *Loch Vennachar*, wrecked on Kangaroo Island, South Australia, in 1905.

Despite the loss of 52 persons, only 4 bodies were ever recovered. The four were identified by Tom Pearce as crew members Reginald Jones and Arthur Mitchell and passengers Mrs Carmichael and Rebecca Carmichael. They were buried above the wrecksite, in coffins made from piano cases washed up from the wreck.

A fifth body, that of Gerald Rolleston, was later found but due to its condition was buried where it was discovered as it could not be moved. Other bodies seen in the water below the surrounding cliffs could not be recovered.

Amazingly, among the debris was a crate containing a fragile full size earthenware Minton Potteries peacock made in 1851, standing over 1.5 metres tall. The item was so fragile that it had been extremely well packed to ensure its safety during the voyage to Melbourne, where it was intended to be displayed in Sydney in 1879 and Melbourne in 1880.

The only damage to the magnificent, full coloured statue was a small break in its beak. It now forms the key part of a display at Warrnambool's Flagstaff Hill Maritime Museum.

The famous Minton Pottery, founded by Thomas Minton, initially made everyday ware, but later generations added the manufacture of artistic pieces which were highly sought after. Today Minton is part of the Royal Doulton Group.

A salvage boat, the *Napier*, visited the site three times with divers, but apart from the peacock found floating in its crate, recovered very little of value. The *Napier* itself was lost while returning to Port Campbell after its third visit.

"The 'Loch Ard' peacock is one of only nine known to have been made by the famous Minton Potteries at Stone-on-Trent. Modelled by the celebrated Italian artist Paul Comolera, the peacock is one of Minton's most important creations. It is fashioned in Majolica, a tin-glazed earthenware which originated in 9th century Islam.
It is thought this is the original to have been modelled by Comolera." FHMV

Eva Carmichael, Captain Gibb and Tom Pearce. *Loch Ard.*

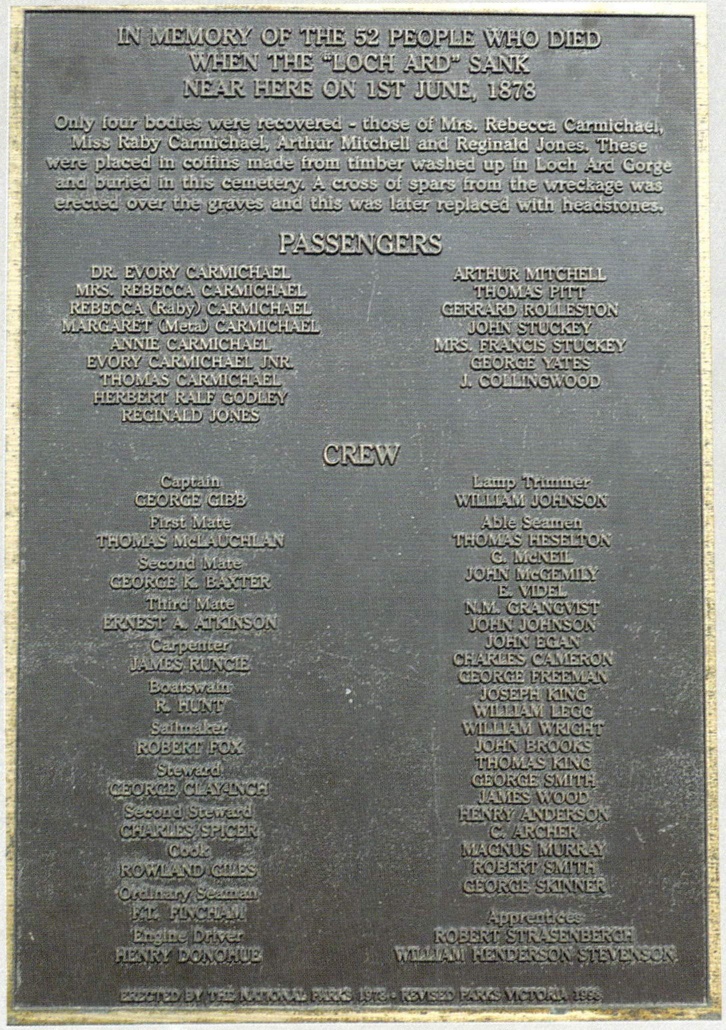

Loch Ard memorial and tombstones, in the most visited graveyard in Australia.

A Minton floor tile, part of the cargo.

Selection of cargo and fittings held by FHMV.

This 10cm porcelain doll most likely belonged to one of Eva Carmichael's younger sisters Margaret (Meta) or Annie.

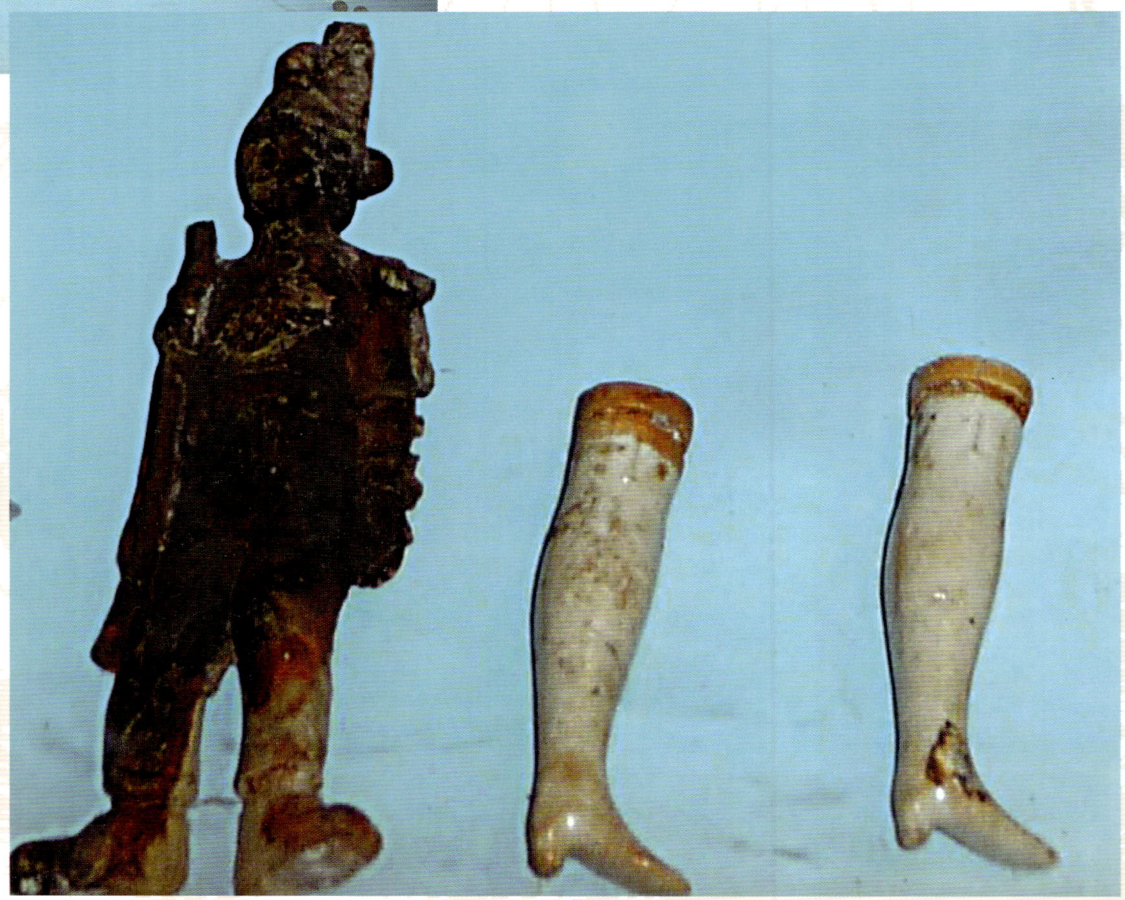

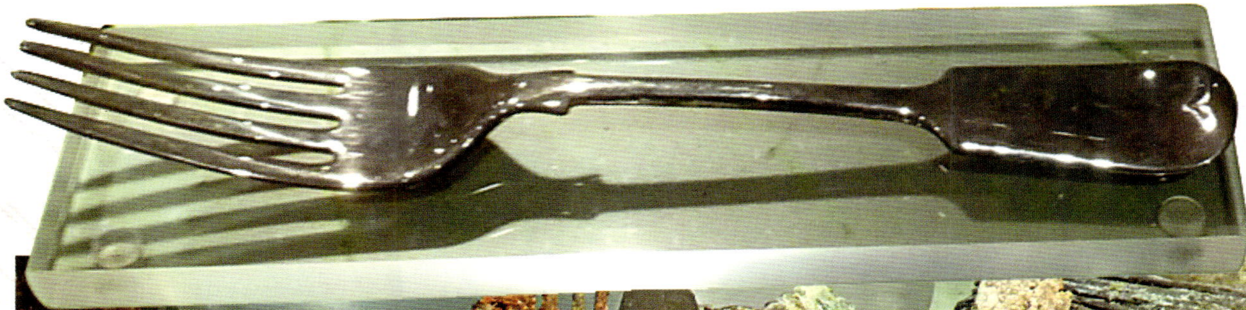

This fork was recovered from the wreck of *Loch Ard* and has been silver plated, demonstrating the appearance of the cutlery prior to the disaster in 1878.

"Although by no means the most luxurious of the ships plying between England and Australia, the *Loch Ard* had carried her 18 passengers in genteel comfort throughout their voage in 1878.
In the evenings, they would have read, played cards or indulged in polite conversation to the soft gas-light thrown by this elegant bracket." FHMV

Schomberg

Described as "the noblest ship that ever floated on water", *Schomberg* was lost on its maiden voyage.

Built in Aberdeen in 1855, the 88 metre long wooden clipper was designed for both comfort and speed, with a massive spread of sail carried on its three masts. Its design included a dining saloon, library and bathrooms for first class passengers, and portholes and ventilation throughout.

Leaving Liverpool on 6 October 1855, *Schomberg* carried 430 passengers, 3,000 tones of cargo and a banner which proclaimed "Melbourne in Sixty Days".

Captain James Nicol "Bully" Forbes had previously established sailing records for the journey, and the ship's owners were pleased and proud to have a master who could achieve this when fast trips meant greater profits.

Light winds around the Equator slowed the giant ship and prevented a record run. Still on Christmas Eve, after a still fast and uneventful trip of 78 days, *Schomberg* was only days from Melbourne when Victoria's south coast was sighted near Cape Bridgewater.

In light winds, *Schomberg* ran onto an unmarked sand spit near Peterborough on 27 December. Sails were taken down, and an anchor let go while a lifeboat was launched to locate the best spot to land passengers. On its return Captain Forbes was advised to wait until the next morning to start the evacuation because of the risk the surf posed to the lifeboats.

The smoke of a steamer was seen at dawn and signal guns fired to attract its attention. The SS *Queen* responded and took aboard all *Schomberg's* passengers. Passenger luggage was collected soon after by another steamer sent from Melbourne by the ship's agents.

Much of the cargo was removed from the doomed ship by steamers from nearby Melbourne until the weather worsened, although 2,000 tons of iron rails remained.

Schomberg broke up on 6 January 1856, scattering timbers and remaining cargo over the nearby coastline. A part of its hull washed up in New Zealand, identified by its distinctive construction which used layers of wood laid diagonally across its frame.

An Inquiry and trial of Captain Forbes resulted in his acquittal. Forbes was not popular as he drove his ships for maximum speed rather than passenger comfort, and during the Inquiry it was established that disgruntled passengers had bribed a crew member to testify against him.

Forbes reputation never recovered and *Schomberg* was to be his last major command.

In 1975 Victorian divers recovered a communion chalice from the wreck, and together with other artefacts, it was displayed at Warrnambool's Flagstaff Hill Maritime Village. Several years later when cleaning the marine growth from the chalice a splendid Brazilian cut diamond ring was found. It is now a feature of the *Schomberg* display.

Contemporary sketch.

In 1975, Peter Ronald and Gary Hayden recovered
an ornate Communion set from the Schomberg,
wrecked at Peterborough in 1855.
It was not until 1980, however, that this
1 carat Brazilian-cut diamond in its gold setting
was discovered embedded in marine concretion
in a lid of the set.
The ring had been in the sea for 120 years.

Community of the Sea

The community of the sea spans generations and oceans. This is evident even in the limited context of the vessels in this book.

Captain James Pearce perished on the *Gothenburg* in Queensland waters, his son Tom Pearce senior was the "hero of the *Loch Ard*" which was lost on the Victorian coast, while his son Tom Pearce junior was a crew member of the ill fated *Loch Vennachar*, lost with all hands off Kangaroo Island, South Australia.

The *Centaur* which rescued the survivors of the German raider HSK *Kormoran*, following its epic battle with HMAS *Sydney* off the Western Australian coast, was the same vessel which, after being converted to a hospital ship, was sunk off the Queensland coast later during World War 2.

The *Yongala* which was the last vessel to see the *Loch Vennachar* in South Australia, was itself lost in Queensland waters.

Watson Bay, where the anchor of the *Dunbar* is on display, is named after Robert Watson, Quartermaster of the *Sirius*, wrecked on Norfolk Island.

The sole survivor of the New South Wales wreck *Dunbar*, James Johnson, rescued the sole survivor of the Paddle Steamer *Cawarra*, Frederick Hedges, off Newcastle.

The widowed Eliza Fraser, survivor of Queensland's *Stirling Castle*, later married Captain Alexander John Greene, the master of the *Mediterranean Packet*, who returned her to England after her rescue.

It was the same William Bligh who accompanied Matthew Flinders while mapping Tasmania who was the Captain of the *Bounty*, taken over by Fletcher Christian and his mutineers. The *Pandora*, sent by the British Admiralty to capture the mutineers, was itself lost off Queensland.

Tom Pearce.

Eliza Fraser.

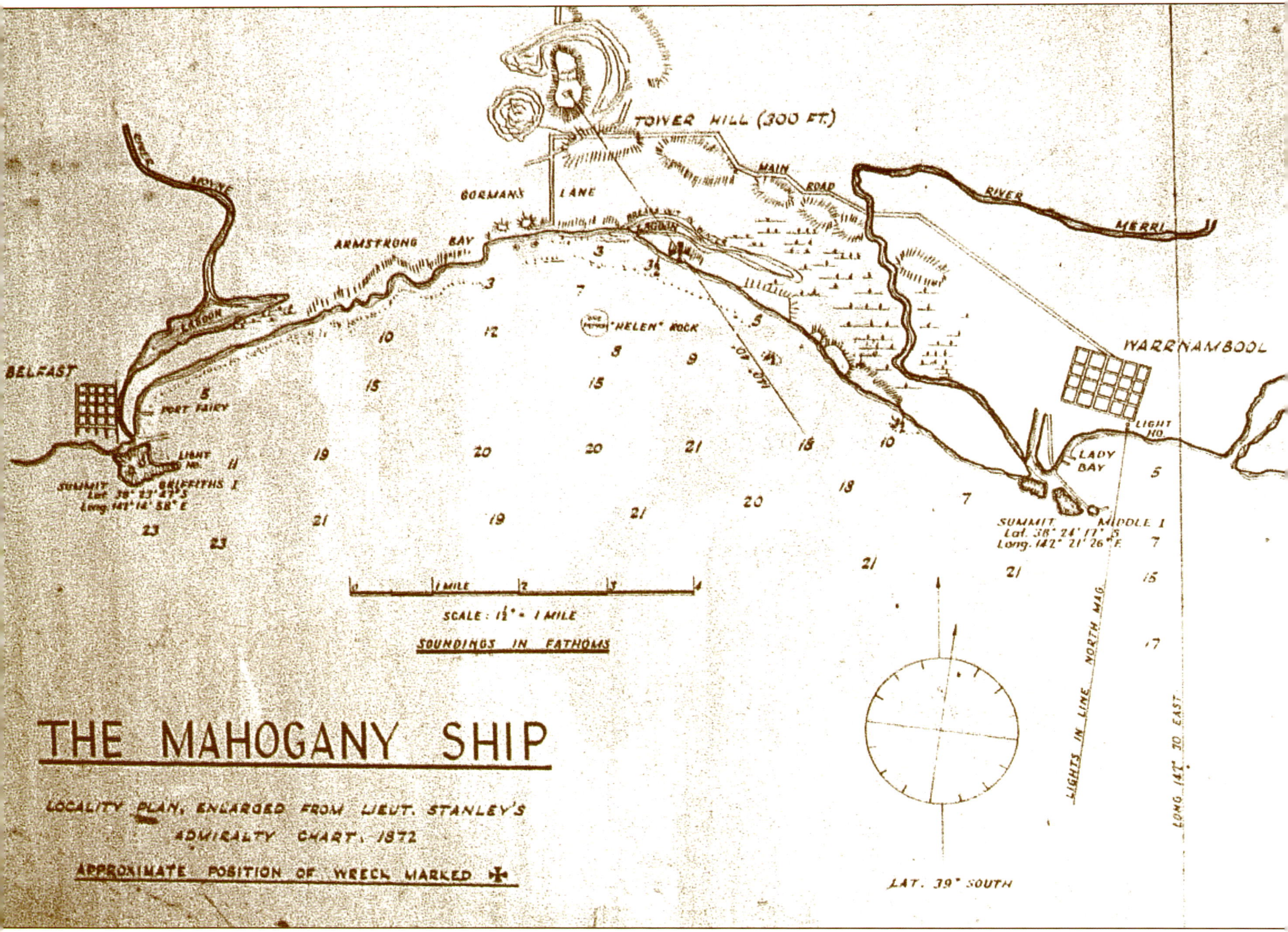

Locality plan from 1872 Admiralty chart FHMV.

Mahogany Ship

The legend of the Mahogany Ship in the sandhills of Victoria's coast has inspired countless searches for it.

First reported in 1836 by whalers, it has been long since covered by shifting sands. Two survivors from a seal hunting trip to the Hopkins River, near Warrnambool came across the remains of a large wooden ship while walking back to their base at Port Fairy after the loss of their boat and a crew member. Port Fairy was the home base of three whaling firms at the time.

Investigating the reported wreck, the captain of the whaling station, Captain John Mills, described it as being constructed of "hard dark timber, like mahogany". The remains lay opposite Tower Hill, half way between Warrnambool and Port Fairy, in sand dunes some distance from the water's edge. Captain Mills went on to become Harbour Master of Port Fairy.

Significant sand movement has occurred continuously in the region causing the disappearance of the Mahogany Ship as it has become known, although it has been sighted many times and its existence is not in any doubt. The most recent reliable sighting of the vessel was in 1880.

As well as being covered by moving sands, it was not unusual for worked timber to be collected and reused, often for housing or sheds, but in other cases simply for firewood by beach users. It has been reported that Alexander Rollo and a friend carried timber to Russell Creek in the 1850's where it was used for furniture. Rollo described the wreck as being four chains (over 80 metres) from the sea, amongst hummocks of sand.

The most accepted theory is that the wreck is of Portuguese origin, with reports of such a loss in 1522. Old Portuguese charts purport to show the southern coast of Australia. Alternative theories put forward the view that the wreck could have been Spanish, French or even Chinese in origin, although the last two are regarded as least likely.

In the same way that Dutch vessels ran into the Western Australian coast after travelling too far east before turning north towards Batavia (now Jakarta) it is not unreasonable to postulate that a ship following a more southerly latitude may have missed that coast and in effect sailed "under" Australia, finally becoming lost upon its southern coast.

Historians vary on their opinions regarding early exploration, but some contend that three Portuguese caravels under the command of Cristovao de Mendonca charted Australia's east coast and the Victorian coastline. The charts, eventually published in 1536 in the Rotz Atlas, end just west of Warrnambool, and one theory is that the two remaining ships turned back after the loss of the third vessel. Arguments continue about whether or not the maps are in fact the Australian coastline.

The Rotz Atlas, also known as "The Boke of Idrography" was produced by Frenchman Jean Rotz and presented to England's King Henry VIII.

Power struggles between Spain and Portugal resulted in great secrecy surrounding journeys of exploration, with outcomes kept undisclosed to maintain any trade advantage that may result from the information. To prevent war Pope Alexander VI, in 1494, divided the globe into two, with Spain and Portugal being restricted to their own half. If in fact explorers from either nation strayed beyond their designated regions they were highly unlikely to publicise the fact that they had done so.

Continual searches have failed to find the remains, which could be under 10 metres of sand. In 1992 the Victorian Government offered a reward of $250,000 for one year for its discovery, but no trace was found. The most recent "professional" search using heavy equipment occurred in 2004.

The discovery of the vessel, should it prove to be of early Portuguese or Spanish origin, would significantly impact upon the accepted history of navigation and exploration of Australia.

Inspired by the story of the Mahogany Ship, a full size sailing replica of a Portuguese or Spanish caravel, the twenty one metre *Raven*, is under construction at Bushfield, near Warrnambool. A walking trail exists along the coast between Warrnambool and Port Fairy, allowing ready access to the region in which the wreck is located. Perhaps one day shifting sands will expose it again.

The Mohogany Ship, painted c. 1860 by Thomas Clark probably from a description.

The Mahogany Ship

Falls of Halladale

After a journey of 102 days from New York, the four masted iron barque *Falls of Halladale* never reached Melbourne, running aground west of the approach to Port Phillip Bay on 14 November 1908.

Its loss only 200 metres from shore resulted in one of the most memorable images of the time, with large crowds gathering on the nearby rocks to look at the stranded ship with its sails still set.

84 metres in length, the Scottish built *Falls of Halladale* was capable of sailing under the most extreme conditions, travelling the world for 23 years in not only the "roaring forties" but the extreme lower latitudes of the fifties.

Its unusual steel masts and wire rigging gave it the strength to deal with heavy conditions far better than ships with wooden masts. Ironically, its loss occurred in calm conditions.

One of the last of the giant windjammers, on its final journey it carried a commercial cargo which included 56,763 slate tiles, 500 sewing machines, iron and over 75,000 litres of oil and benzine.

In calm but misty conditions, Captain Thompson believed his ship was fifteen kilometres off the southern Victorian coast. By the time he realised this was not the case it was too late to turn the vessel around and the *Falls of Halladale* went aground on rocks near Peterborough.

The ship struck within two minutes of land being sighted, and immediately began taking water.

Two lifeboats were lowered, and not being able to go ashore due to rocks, the crew rowed almost eight kilometres to a beach in the Bay of Islands.

Witnessing the stranding, Peterborough's Postmistress raised the alarm. As a result of the large number of shipping accidents in the region, she had been provided an observation telescope by the Government to keep a watch for vessels in distress.

The rocket crew from Port Campbell arrived, but were not required as the ship's crew had safely reached shore. No lives were lost, however the *Falls of Halladale* became a complete loss.

The equipment, although not needed on this occasion, was used to rescue people by firing rockets carrying safety lines to ships so that stranded passengers and crew could be brought ashore, often through heavy seas.

Much of the cargo was recovered, although a significant amount was lost as the ship was broken up by the sea in the weeks following its stranding. Explosives used during salvage operations caused some of the cargo of oil to be spilled and washed ashore.

Twenty two thousand tiles were recovered between 1974 and 1986 and used in Warrnambool's Flagstaff Hill Maritime Village, where they can be seen on the roofs of restored buildings.

Captain Thompson was found guilty of gross misconduct by a Court of Marine Inquiry, and had his Certificate cancelled for six months.

"Section of Name Plate — Falls of Halladale
– the ship's name appeared in large brass letters across the stern and on both port and starboard bows." FHMV

Paper roll, clock mechanism and anchor from *Falls of Halladale*.

TASMANIA

Cataraqui

The emigrant vessel *Cataraqui* has the dubious honour of the most lives lost in Australia's maritime history during peacetime.

The Canadian built barque had been purchased to carry emigrants from England to Melbourne, and leaving Liverpool on 20 April 1845 carried 367 emigrants, 41 crew and Captain Christopher Finlay.

Sixty one families were aboard and almost half the passengers were children under the age of 14 years.

Passenger numbers increased with five babies born during the trip, but six others died. A crew member was also lost overboard, so by the time *Cataraqui* approached the end of its voyage it carried 408 persons.

Stormy weather for the last two weeks of July gave the ship speed, but made accurate navigation difficult, as regular sightings of the sun were necessary to determine its position.

Apart from the discomfort of the storms, the trip was otherwise uneventful until *Cataraqui* entered Bass Strait on the morning of 4 August, where bad weather was encountered. During the gale *Cataraqui* struck rocks off the south west coast of King Island at 4.30 am.

Captain Finlay believed his ship to be eighty kilometres south west of Portland, on a course for Port Phillip Heads. In fact he was nearly 100 kilometres further south.

Half those aboard the doomed ship were trapped below decks and drowned, with many of the remainder on deck washed overboard by the heavy seas. Lifeboats could not be launched. By evening the ship had broken in two, with one half sinking. The remaining half was smashed by the seas over the next day or so, taking most still alive upon it to their deaths as it disappeared.

Only nine persons managed to survive, eight crew and one passenger, Samuel Brown. Brown's wife and four children were amongst those lost.

An account by one of the survivors, Chief Mate Thomas Guthery (Guthrie in some reports), states in part *"About half past four o'clock am, the tempest still continuing, rain falling in torrents, and the darkness of the night such as not to allow any object to be seen within a few hundred yards of the vessel, the ship struck, going at the rate, I suppose, of about seven knots through the water. ....striking the third time, she struck and fell over on her larboard side. At this time the sea commenced breaking on board, the passengers at once rushing on deck, and crying for assistance. ....The sea by this time was breaking on deck, and sweeping everything away, with part of the crew and passengers.....At daybreak, we observed the land ahead of the vessel, I should say about 150 yards, and could likewise see that the vessel was breaking up aft. The sea was breaking at this time over the vessel, and washing away at every sea some of the passengers and crew."*

The survivors were found by sealers, and remained in their care until 7 September when picked up by the cutter *Midge*. In the meantime people in Melbourne awaiting family and friends were becoming anxious at the delay in the ship's arrival. The appearance of the *Midge* with the few survivors saw their worst fears realised.

In total 342 bodies were recovered and buried together in five large graves, and a number of smaller ones, generally in the vicinity in which they were found.

Some reports indicate that *Cataraqui* carried 369 emigrants, resulting in varying accounts of either 399 or 401 lives lost. Whatever the actual figure, the loss was only surpassed by ships sunk in wartime.

Cataraqui was named after the city of Cataraqui, now called Kingston, in Ontario, Canada.

The wreck of the *Cataraqui* — drawn by one of the few survivors of the wreck, Ordinary Seaman Peter Johnson.

Mass grave for 245 bodies.

Field naturalist excursion to King Island, including monument at Cataraqui Point.

Diver with porcelain and glass bottles on *Sydney Cove* site.

Sydney Cove

Believed originally to be the Indian ship *Begum Shaw*, it is significant that the quite small, 30 metre, *Sydney Cove* set out from the Bay of Bengal on 10 November 1796 to trade with the new colony of Sydney.

After a three month journey from India, travelling down the west coast and south of the continent, *Sydney Cove* was leaking to the extent that its master, Captain Guy Hamilton, beached the vessel on what is now Preservation Island on 9 February 1797.

Severe gales in December damaged the bow of the ship. The repairs made at the time were further strained in January during another gale, forcing the continuous operation of the pumps to keep afloat. Five of the Indian crew died following the second incidence of damage, with accounts showing this to be from both scurvy and exhaustion. Severe conditions while in Bass Strait on 8 February sealed the ships fate.

As rescue was unlikely due to the remoteness of the Island and the limited number of ships in the region, a decision was made for seventeen of the crew to sail a longboat to Port Jackson for help. They left on 27 February but were wrecked on the Victorian coastline. All seventeen survived, but now faced a 600 kilometre overland journey to Sydney. Only three survived the trek, being picked up by a fishing boat and arriving at Port Jackson on 16 May.

The balance of the crew remained behind and continued to unload cargo from the beached vessel over coming weeks. Four more crew died during the unloading. *Sydney Cove's* cargo consisted in the main of 31,500 litres of spirits, mostly rum, about 60% of which was salvaged. Cargo also included bottled wine, shoes and leather skins..

It became necessary for Captain Hamilton to move casks of spirits to a nearby island to prevent the crew drinking them, leading to it being named Rum Island.

On 10 June the small rescue boats *Francis* and *Eliza* arrived on Preservation Island, taking aboard both cargo and the stranded

survivors. Six of the original crew stayed behind to guard the remaining cargo that could not be taken on the two small vessels.

The *Eliza* was lost on the return journey to Sydney, with its crew and eight of the *Sydney Cove's* crew who were aboard after being rescued.

The *Francis* made two more trips picking up the remaining crew and cargo and completing its final voyage on 3 March, with some of the crew that left India having spent a year on Preservation Island. Aboard the third voyage of the *Francis* was Lieutenant Matthew Flinders who was responsible for mapping the area.

Remains of the *Sydney Cove* were located again in 1977, resulting in extensive underwater and land based archaeological surveys over coming years.

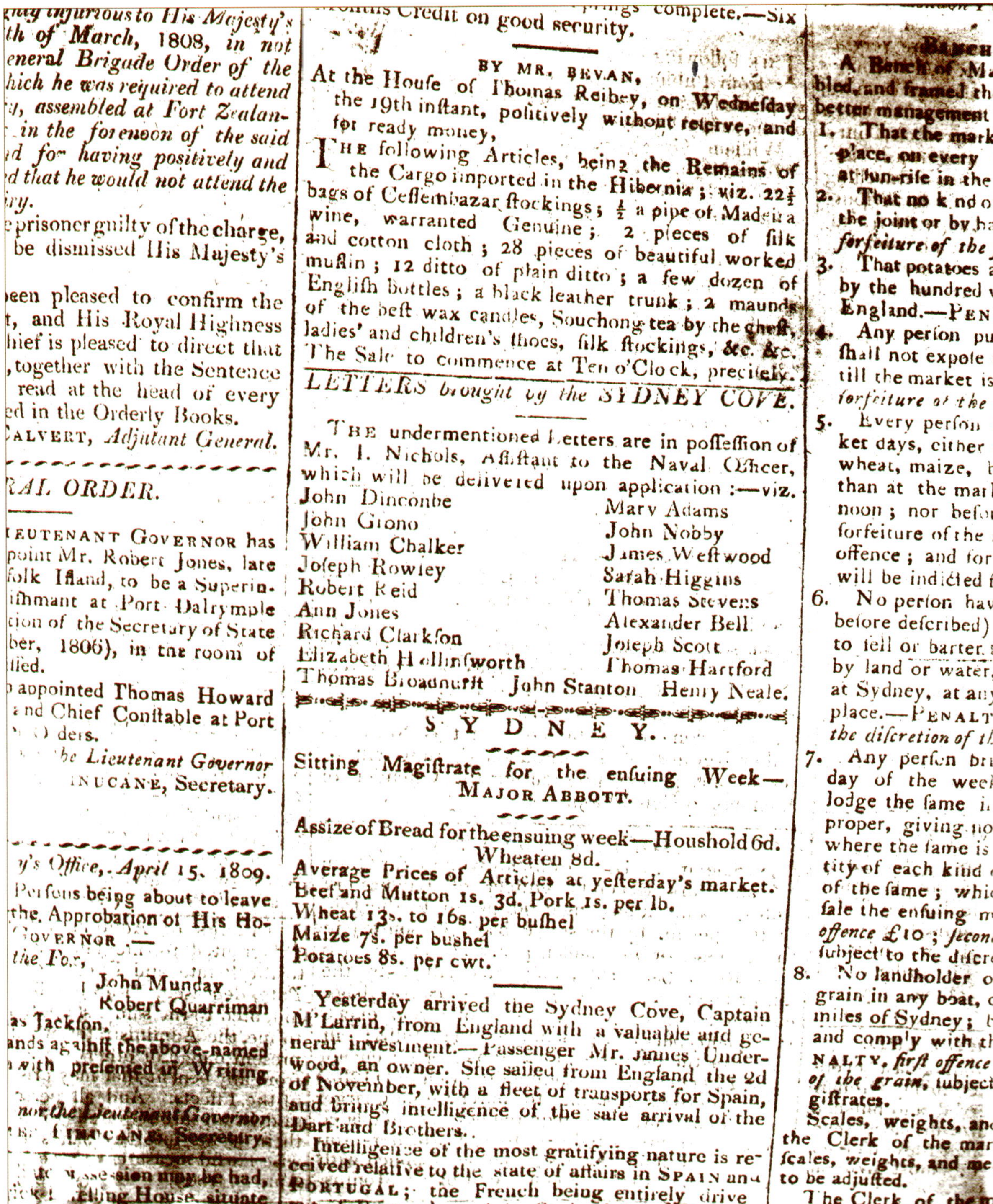

ing injurious to His Majesty's
th of March, 1808, *in not*
eneral Brigade Order of the
hich he was required to attend
, assembled at Fort Zealan-
in the forenoon of the said
d for having positively and
d that he would not attend the
ry.
prisoner guilty of the charge,
be dismissed His Majesty's

een pleased to confirm the
t, and His Royal Highness
hief is pleased to direct that
, together with the Sentence
read at the head of every
d in the Orderly Books.
ALVERT, *Adjutant General.*

AL ORDER.

IEUTENANT GOVERNOR has
point Mr. Robert Jones, late
folk Island, to be a Superin-
ishmant at Port Dalrymple
tion of the Secretary of State
ber, 1806), in the room of
lied.
appointed Thomas Howard
and Chief Constable at Port
ders.
he *Lieutenant Governor*
INCANE, Secretary.

y's Office, April 15. 1809.
Persons being about to leave
the Approbation of His Ho-
GOVERNOR.—
the Fo,
John Munday
Robert Quarriman
as Jackson.
ands against the above-named
with presented in Writing
nor the Lieutenant Governor.
INCANE, Secretary.

...nths Credit on good security.

BY MR. BEVAN,

At the House of Thomas Reibey, on Wednesday the 19th instant, positively without reserve, and for ready money,

THE following Articles, being the Remains of the Cargo imported in the Hibernia; viz. 22½ bags of Cessembazar stockings; ½ a pipe of Madeira wine, warranted Genuine; 2 pieces of silk and cotton cloth; 28 pieces of beautiful worked muslin; 12 ditto of plain ditto; a few dozen of English bottles; a black leather trunk; 2 maunds of the best wax candles, Souchong tea by the chest, ladies' and children's shoes, silk stockings, &c. &c. The Sale to commence at Ten o'Clock, precisely.

LETTERS brought by the SYDNEY COVE.

THE undermentioned Letters are in possession of Mr. I. Nichols, Assistant to the Naval Officer, which will be delivered upon application:—viz.

John Dinconbe — Mary Adams
John Giono — John Nobby
William Chalker — James Westwood
Joseph Rowley — Sarah Higgins
Robert Reid — Thomas Stevens
Ann Jones — Alexander Bell
Richard Clarkson — Joseph Scott
Elizabeth Hollinsworth — Thomas Hartford
Thomas Broadhurst — John Stanton — Henry Neale.

SYDNEY.

Sitting Magistrate for the ensuing Week—MAJOR ABBOTT.

Assize of Bread for the ensuing week—Houshold 6d. Wheaten 8d.

Average Prices of Articles at yesterday's market. Beef and Mutton 1s. 3d. Pork 1s. per lb. Wheat 13s. to 16s. per bushel Maize 7s. per bushel Potatoes 8s. per cwt.

Yesterday arrived the Sydney Cove, Captain M'Larrin, from England with a valuable and general investment.—Passenger Mr. James Underwood, an owner. She sailed from England the 2d of November, with a fleet of transports for Spain, and brings intelligence of the safe arrival of the Dart and Brothers.

Intelligence of the most gratifying nature is received relative to the state of affairs in SPAIN and PORTUGAL; the French being entirely drive out of Portugal; and in Spain defeated...

A Bench of Ma...
bled, and framed the...
better management...
1. That the mark...
place, on every...
at Sun-rise in the...
2. That no kind of...
the joint or by ha...
forfeiture of the f...
3. That potatoes a...
by the hundred w...
England.—PEN...
4. Any person pu...
shall not expose t...
till the market is...
forfeiture of the...
5. Every person i...
ket days, either v...
wheat, maize, b...
than at the mark...
noon; nor befor...
forfeiture of the a...
offence; and for t...
will be indicted fo...
6. No person havi...
before described)
to sell or barter t...
by land or water,
at Sydney, at any
place.—PENALT...
the discretion of th...
7. Any person bri...
day of the week...
lodge the same i...
proper, giving no...
where the same is...
tity of each kind o...
of the same; whic...
sale the ensuing m...
offence £10; secon...
subject to the discre...
8. No landholder o...
grain in any boat, c...
miles of Sydney; b...
and comply with th...
NALTY, *first offence*
of the grain, subject
gistrates.
Scales, weights, and
the Clerk of the mar...
scales, weights, and me...
to be adjusted.
The Clerk of the m...

Neva

Sailing for Sydney from Cork, Ireland on 8 January 1835, *Neva* carried 150 female convicts and their 33 children, as well as another 31 free women and children.

Surgeon Superintendent John Stephenson was responsible for the medical welfare of the convict women and crew, and the journey was undertaken with only the loss of three lives until reaching the Australian coastline. There had also been one birth.

Under the command of Captain Benjamin Peck, the *Neva's* passengers and crew totalled 239 when on 13 May 1835 the ship struck a reef off King Island.

Having calculated the ship's position on 12 May as being 150 kilometres to the west of King Island, the Captain set course to sail 22 kilometres north of the Harbinger Reefs through Bass Strait to Sydney.

At 2.00am of the 13th a more northerly course was set after sighting land 15 kilometres ahead. Three hours later breakers were seen by the lookout. The ship was turned, but it was too late, striking a submerged rock and losing its rudder. The bow then struck the main reef, and the ship quickly filled with water.

Panic set in as the ship was overwhelmed by the heavy seas, with 200 women and children gathering on the main deck. As they huddled at either end of the ship the stores were broken into.

The effect of rum being carried on board as cargo is reported as being a significant factor in so few surviving, with many incapable of looking after themselves.

Lifeboats could not be launched as the ship was side on to the reef and breaking up. As the ship broke in half, two sections remained afloat and survivors scrambled upon these rafts, which drifted onto King Island.

Twenty two survivors made it to shore, but only fifteen survived the first night. 224 lives had been lost.

Timber from the wreck was used to build a rough shelter, and cargo washed up included casks of salt pork and flour, giving the survivors some hope. Ninety five bodies were found and buried, but there were no more survivors.

Captain Peck and the fourteen survivors were found two weeks later by the crew of the small cutter *Tartar* which had also been lost 50 kilometres south, and were taken to the hut of John Scott, a sealer, and his aboriginal wives and family.

On 15 June the *Neva* and *Tartar* survivors were rescued by the *Sarah Ann*, a vessel chartered by the *Tartar's* owner to search for his missing boat. After her harrowing experience, one woman, Rose Hyland, refused to board the rescue ship and had to be forcibly carried aboard.

Three *Neva* survivors who had been off hunting when the rest were rescued were subsequently picked up by the *Charlotte*, a Government cutter sent out for that purpose.

The *Neva* was a 32 metre barque built in 1813 and converted to a convict carrier in the early 1830's.

Captain Peck claimed to have hit an uncharted reef, although it is more likely that after failing to take proper account of current and local conditions *Neva* hit the known Harbinger Reefs. The Captain was exonerated by a subsequent Inquiry.

A full and Horrifying Account of a most dreadful

SHIPWRECK

Of the Neva, Convict Ship, Which sailed from Cork, for New South Wales, in May last, containing 241 Souls on Board, consisting of 150 female Convicts, 9 free Women, and 55 Children, all of whom perished except 6; and 9 of the Crew; with an Account of the dreadful Sufferings of those who were saved, while staying on a desert Island, being altogether one of the most heart-rending Accounts of Human Suffering which has occurred for a very long Period.

South Australia

Admella

> "....the brave men who risked their own lives to save the lives of their fellow creatures - the noble, daring, heroic band who ventured upon the breaking crests of those impending billows - who plunged into the seething troughs of that foaming sea - who, repulsed again and again, ventured again and again, at the renewed risk of their own lives to save the lives of strangers."
>
> Adelaide "Advertiser" Monday August 15, 1859 reporting the *Admella* rescue efforts.

The tragic loss of the steamship *Admella* in 1859 has not been paralleled by any other wreck in South Australia. The massive loss of life and the horrendous circumstances of the sinking have left the *Admella* a permanent place in our maritime history.

Leaving Port Adelaide on Friday 5 August, the *Admella* set course for Victoria with a cargo of passengers, copper, flour and seven horses. A total of 113 people were on board.

Early that afternoon, while in Backstairs Passage, one of the racehorses, Jupiter, fell in its box. Course was altered for an hour to reduce the effect of the ocean swell while the horse was put back onto its feet. Normal course was resumed and Captain McEwan continued with little effect on his schedule, until the early hours of the following day, believing the *Admella* to be a safe distance from shore.

At 5 a.m. Saturday, the 6th, the *Admella* struck a reef on Carpenter's Rocks, two kilometres off shore. The swell lifted the vessel higher onto the rocks, where it lay on its port side. In just over ten minutes the *Admella* had broken into three sections. Within hours the mid section had slid down off the rocks into deep water.

The tragedy, however, was only now just beginning.

Several lives were lost almost immediately as the vessel broke up, throwing people and rigging into the angry sea. One of the lifeboats was also crushed by a falling funnel. Signal rockets were fired, but were not seen on the thinly populated coast. The lighthouse keepers at Cape Northumberland were over twenty kilometres away.

That morning the *Admella's* sister ship, *Havilah*, passed in the distance without seeing the stricken vessel. Several attempts to get ashore on floating timber failed, and resulted in further lives being lost. An attempt to secure one of the ship's boats seen floating upside-down nearby saw another life lost. Hunger , thirst and cold were also now becoming a problem as night fell.

The steamship *Bombay* passed close to the wreck during the night, but did not see it, as there were no lights or rockets to signal with. The next morning fifteen men crossed from the bow section to the stern, although a further five lives were lost, including two children. Within hours the bow section broke up, drowning those men, women and children who remained.

A raft was built by sailors during the day, and after several hours in the rough seas, it reached shore with the first of the survivors, who were able at last to take word of the disaster to the keeper at the Cape Northumberland lighthouse.

The rescue effort was then seriously started with details of the wreck being advised to the Mount Gambier Post Office, which was able to advise the appropriate authorities. The nearest lifeboat, however, was in Portland, three hundred kilometres away.

The steamer *Corio* and two fishermen from Robe reached the wreck before the lifeboat but were prevented by the heavy seas from providing immediate assistance. By this time over fifty lives had already been lost. The torment of those still surviving, with minimal food, water or shelter, trapped in enormous seas, is frightening to imagine.

An attempt by the *Corio's* boat to reach the wreck on Thursday, the 11th, proved unsuccessful. The steamers *Lady Bird* and *Ant* arrived on that day, the former towing the Portland lifeboat. Early on Friday the lifeboat tried to reach the wreck, but huge seas broke the rudder and several oars. Many rescue attempts were similarly unsuccessful, with mountainous seas and rugged reefs combining to damage rescue craft and injure rescuers.

On Saturday, the 13th, the lifeboat, and the *Corio's* boat managed to secure a line to the *Admella's* remains, and eventually three survivors, including Captain McEwan, were rescued. The final nineteen survivors were rescued, eight days after being wrecked, on Sunday 14th.

In all, eighty six lives were lost in this tragedy, including fifteen children.

The week long rescue efforts had been of intense interest to all in the colonies, with normal business being effectively suspended. The tragedy was so all pervading that normal business was virtually impossible, and The Advertiser office was stormed by people desperate for information.

"Many a heart will bleed to think of what must have been suffered by those who so long clung to the iron fragments of the wreck, beating on those hard rocks, amidst the furious lashing of the waters, for half the time without sun or moon to guide, and without a hand to help them. Many will think the less terrible was their fate who went down in the fore part of the vessel when it struck, and parted, and sunk, than the fate of those who succeeded in gaining the after-portions of the vessel, only to die lingering deaths by hunger, exhaustion, and fatigue, after seeing their fellow-sufferers one by one disappear" reported The Advertiser.

Substantial funds were collected to assist the survivors and families of those lost. In addition awards for bravery were presented to those who participated in the rescue efforts.

The *Admella* was an iron steamship of 500 tons, built in Glasgow in 1857, measuring 59 metres in length. The name *Admella* was derived from the initial letters of **AD**-elaide, **MEL**-bourne and **LA**-unceston.

The famous lifeboat used in the rescue is on display in Portland.

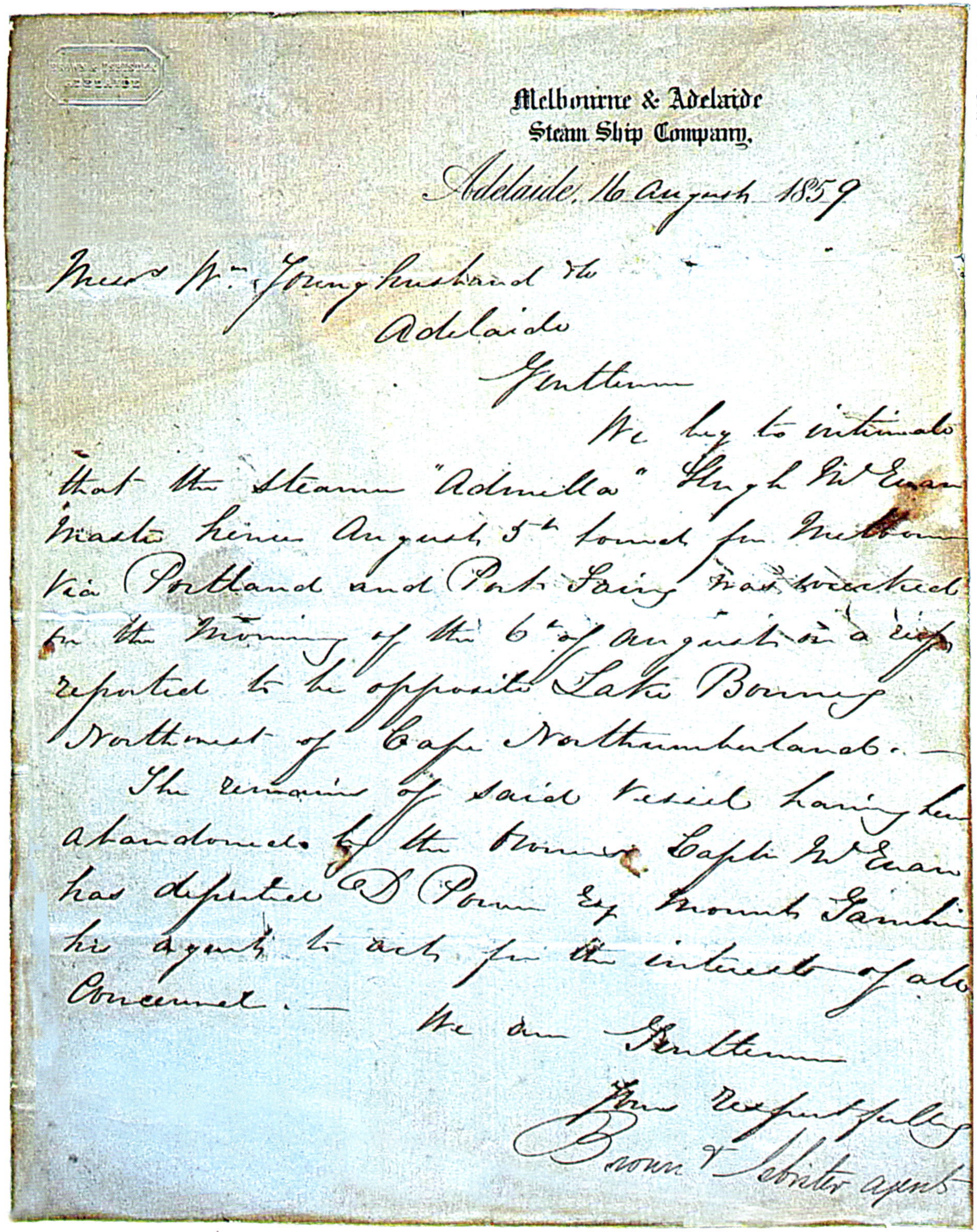

Melbourne & Adelaide
Steam Ship Company.

Adelaide, 16 August 1859

Messrs Wm Younghusband & Co
Adelaide

Gentlemen

We beg to intimate that the Steamer "Admella" Hugh McEwan Master hence August 5th bound for Melbourne via Portland and Port Fairy was wrecked on the morning of the 6th of August on a reef reported to be opposite Lake Bonney Northwest of Cape Northumberland.—

The remains of said vessel having been abandoned by the owners Captn McEwan has deputed A D Power Esq Mount Gambier his agent to act for the interests of all concerned.—

We are Gentlemen

Yours respectfully
Brown & Webster agents

Official notification of the loss of the *Admella*.

Medal awarded to Peter Smith

Luxury first class cabin fitting.

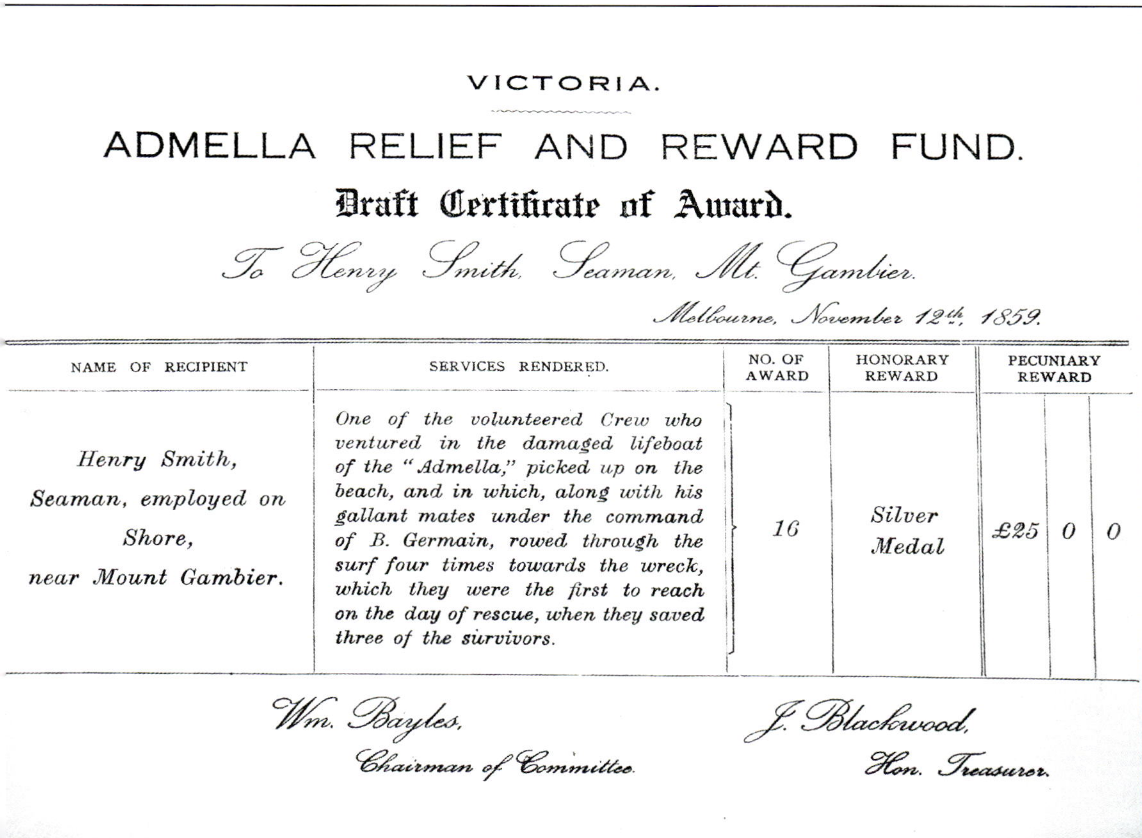

VICTORIA.

ADMELLA RELIEF AND REWARD FUND.

Draft Certificate of Award.

To Henry Smith, Seaman, Mt. Gambier.

Melbourne, November 12th, 1859.

NAME OF RECIPIENT	SERVICES RENDERED.	NO. OF AWARD	HONORARY REWARD	PECUNIARY REWARD		
Henry Smith, Seaman, employed on Shore, near Mount Gambier.	*One of the volunteered Crew who ventured in the damaged lifeboat of the "Admella," picked up on the beach, and in which, along with his gallant mates under the command of B. Germain, rowed through the surf four times towards the wreck, which they were the first to reach on the day of rescue, when they saved three of the survivors.*	*16*	*Silver Medal*	*£25*	*0*	*0*

Wm. Bayles,
Chairman of Committee.

J. Blackwood,
Hon. Treasurer.

Admella memorial at Cape Banks.

Clan Ranald

The *Clan Ranald* was a most unusual ship in appearance.

Built in 1900, the 108 metres long, 3596 ton vessel was a turret ship. This design featured a shoulder near the top of its hull, so that the deck was narrower than its actual beam. A cross section of the hull gave the appearance of a vase, rather than the traditional "V" shape.

Storage of grain was facilitated by the design, but it also had the advantage of reduced operating costs as payment for passage through the Suez Canal was based on a calculation which included the deck width. *The Clan Ranald's* deck was officially narrower than its width at the "shoulder" of the vase shape.

In January 1909 the *Clan Ranald* had taken on a load of 5000 tons of wheat and flour at Port Adelaide for South Africa. Prior to departure over 600 tons of coal were loaded, causing a list to starboard. This problem was overcome by the movement of 50 tons of coal. An inspection by a Lloyd's surveyor during loading confirmed that the cargo was properly stowed.

The *Clan Ranald* left Port Adelaide on the morning of Sunday 31 January after being towed from the Port River by the tug *Eagle*. The *Eagle's* crew reported that she was again listing slightly to starboard.

Soon after noon the vessel was in the vicinity of Troubridge Lighthouse, steaming at just over 8 knots. At 2 p.m. she suddenly heeled to starboard at an angle of 45 degrees. Attempts were made to free the starboard lifeboats, but this side of the ship was underwater and the lifeboats were entangled in their tackle.

The weather conditions deteriorated and for the next few hours the *Clan Ranald* drifted helplessly towards Troubridge Hill. Soon after 8 p.m. an anchor was put down when the ship was approximately one kilometre from shore.

Distress rockets were fired and hope of rescue was high when the lights of another ship were seen. Unfortunately this vessel, the *Uganda*, did not come to assist, with her Captain later claiming not to have seen the distress signals.

Lifebelts were distributed to all the crew and work began to build wooden rafts from the limited amount of available timber. Suddenly, at 10 p.m. the *Clan Ranald* rolled over and sank, throwing the crew of 63 into the sea. Only 23 survived.

A farmer had noticed that the ship was in trouble during late afternoon and went to Edithburgh to alert the Harbourmaster. However, as he was at Sunday evening church service, the report was not received by him until 8 p.m. Various shore based activities were put into effect, but by the time the they reached Troubridge Hill all they could do was help the few survivors and collect the battered bodies.

A Marine Board of Inquiry was set up in Adelaide the following week, but was unable to establish the reason for the disaster. An inspection by a diver found no evidence of the hull being damaged. Shifting of cargo was also dismissed as a possible reason for the listing as an inspection prior to departure had confirmed that this was properly stowed.

The steel hulled, 3596 ton steamship now lies in 20 metres of water, one kilometre from the cliff upon which her anchor is mounted as a memorial. The *Clan Ranald* wreck is semi intact, resting upside down, with the huge boiler near the middle of wreckage that is still identifiable as a ship.

Survivors were brought to Adelaide, although the Lascar crew were treated as illegal immigrants under the "White Australia" policy that existed at the time and forced to leave less than a fortnight after the wreck on another ship of the Clan Line.

The white officers and coloured crew are buried in separate graves in the Edithburgh cemetery, although only the officers' names were listed. Relics of the vessel are located in the Edithburgh Museum.

In January 2009 the centenary of the *Clan Ranald's* loss was commemorated in Edithburg, with a new plaque unveiled in the cemetery listing the names of the Lascar crew.

The *Clan Ranald* is a rare example of a turret deck steamship.

The original 1909 funeral procession at Edithburg. It was to be 100 years before the names of the Lascar crew were commemorated.

Compass holder.

The *Clan Ranald's* unusual turret deck design is shown here..

STAR OF GREECE
BELFAST

Star of Greece

Fear of a Russian invasion led to the *Star of Greece* being in Adelaide in 1888. It was one of two ships which brought large cannons to the colony after Russian ships had been sighted in South Australian waters several years earlier. The cannon were never fired in anger but perhaps served their purpose. The Russians never did invade South Australia!

The *Star of Greece* was a 3 masted iron ship of 1227 tons, built at Belfast in 1868 and owned by Corry & Co. Measuring 69 metres in length she was of particularly fine lines and was compared by many to a yacht, rather than a cargo vessel, so fine were her features. Most of her 25 voyages were between London and Calcutta and on one of these in 1877 the *Star of Greece* equalled the speed record set by the famous clipper *Cutty Sark*.

The visit to Adelaide was the *Star of Greece's* third to Australia and the first under the command of Captain Harrower. On 12 July 1888 the ship, with a full cargo of 16000 bags of wheat, had completed all arrangements for the return voyage to the United Kingdom, including clearance by Port Adelaide Customs. A storm which had been raging for two days had already delayed her departure.

Despite the storm, and perhaps because of the risk of crew desertions as well as a desire for a fast return journey, the Captain gave the order to raise anchor and go late in the afternoon. The weather was relatively calm at the time of departure but had come up again by the time the *Star of Greece* was just past Semaphore. Two hours after departure, sail was shortened in order to delay the vessel's arrival in Backstairs Passage until daylight.

At 2.30am on the 13th July the shore was sighted and the port anchor let go with 60 fathoms of chain. The starboard anchor could not be released. The anchor did not hold and the ship drifted helplessly towards shore. At 3am the ship's bow struck. By 4am the *Star of Greece* was totally aground at Pt. Willunga, sea breaking over her. Within two hours her back was broken and masts shattered. An attempt to launch a lifeboat was unsuccessful, with the huge seas destroying it against the ship.

The wrecked ship and her crew were only 200 metres from shore. During the morning attempts were made by some of the crew to swim ashore, but they were either swept back or drowned. At 1p.m. Hazeland, the First Mate, made it to shore and safety. In all ten of the crew of twenty seven survived. Many of those lost were mutilated by wreckage while attempting to swim to shore.

The wreck was sighted from shore at 7a.m. and rescue efforts were underway at the same time as the crew tried to save themselves. These shore based rescue efforts were severely limited, however, as the rocket apparatus needed for the rescue had been removed years earlier by a Government cost cutting exercise. It is likely the whole crew could have been saved had the equipment been available. The savings in maintenance costs to the Government were in the order of six pounds a year.

Although people at Pt. Willunga knew of the wreck at 7am, it was over two hours later before the message could be got to Adelaide or Normanville because of the hours of operation of the Telegraph Station at Willunga. Rocket gear did eventually arrive at Port Willunga at about 5pm but by this time it was too late to be of use as nobody remained on the wreck. The steam vessels *Defiance* and *Yatala* attended the wreck site on the 13th but were not able to assist in rescue operations because of the sea conditions and the location of the *Star of Greece* so close to shore.

A political furore erupted following the wrecking. The lack of properly maintained rescue equipment along the coast was a major issue, as was the decision not to send the warship *Protector* to the wrecksite to assist with rescue operations. The Marine Board was found negligent by a Parliamentary Select Committee, and responsibility for lifesaving operations was given to the Navy, in 1888 still under the control of State Governments.

Salvage operations began soon after the sale of the wreck, and in August the main anchor was recovered. It was found that only 40 fathoms of chain had been released just prior to grounding, not 60 fathoms as thought. In addition the recovery crew discovered that the chain had wrapped itself around the anchor, preventing it from holding.

The *Star of Greece* now rests in 6 metres of water and is easily accessible, with part of a mast visible from the shore at low tide. The figurehead and other relics are on display in the South Australian Maritime Museum at Port Adelaide.

Dr. McGowan (*front right*) and Mrs Bowering with fellow rescuers.

Brass fire bucket with letter 'F'. One of 12 making up the ship's name.

Port Willunga cemetery.

Underwater plaque at wreck site, placed by the Department of Environment and Heritage.

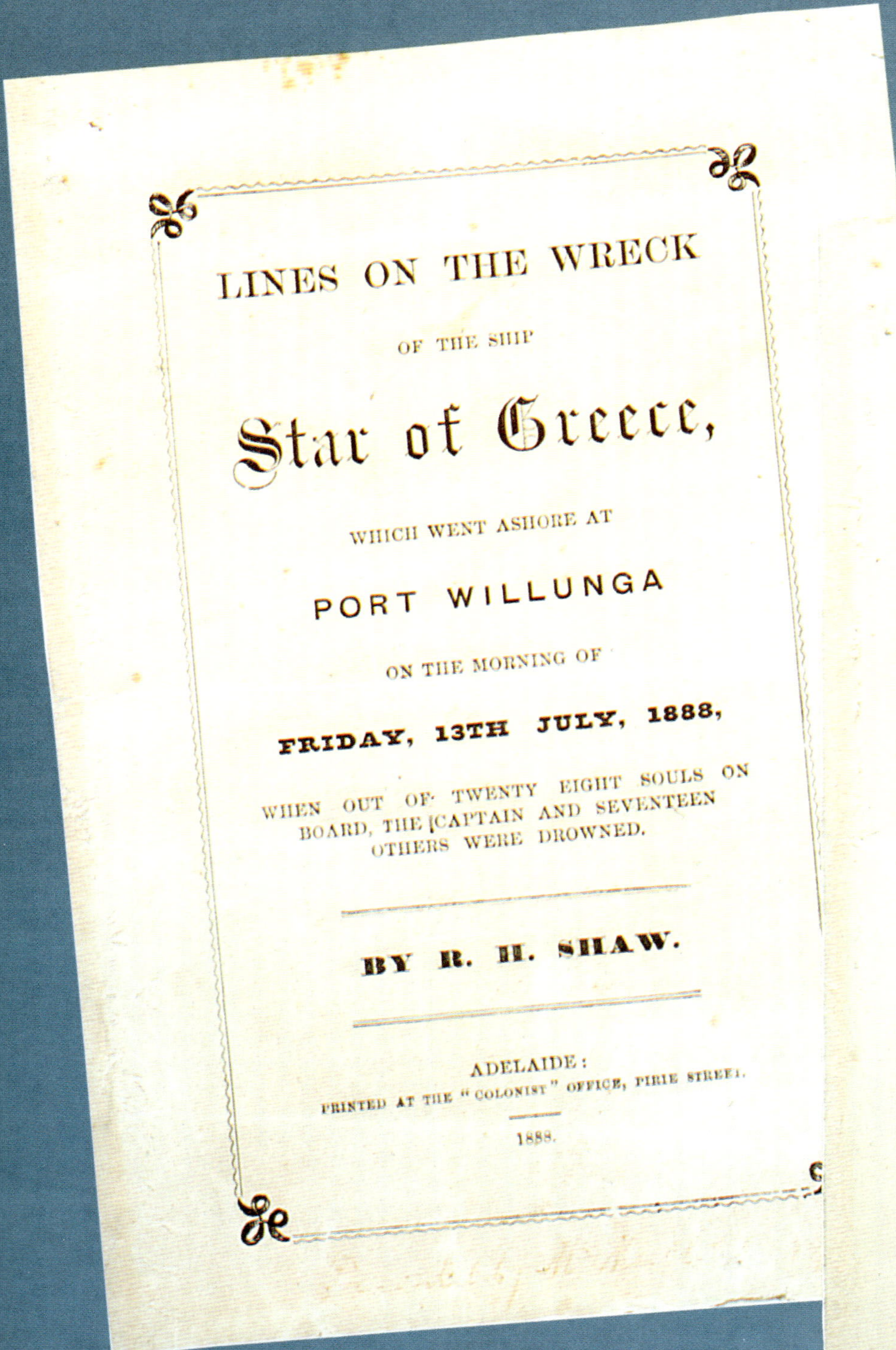

LINES ON THE WRECK

OF THE SHIP

Star of Greece,

WHICH WENT ASHORE AT

PORT WILLUNGA

ON THE MORNING OF

FRIDAY, 13TH JULY, 1888,

WHEN OUT OF TWENTY EIGHT SOULS ON BOARD, THE CAPTAIN AND SEVENTEEN OTHERS WERE DROWNED.

BY R. H. SHAW.

ADELAIDE:
PRINTED AT THE "COLONIST" OFFICE, PIRIE STREET.

1888.

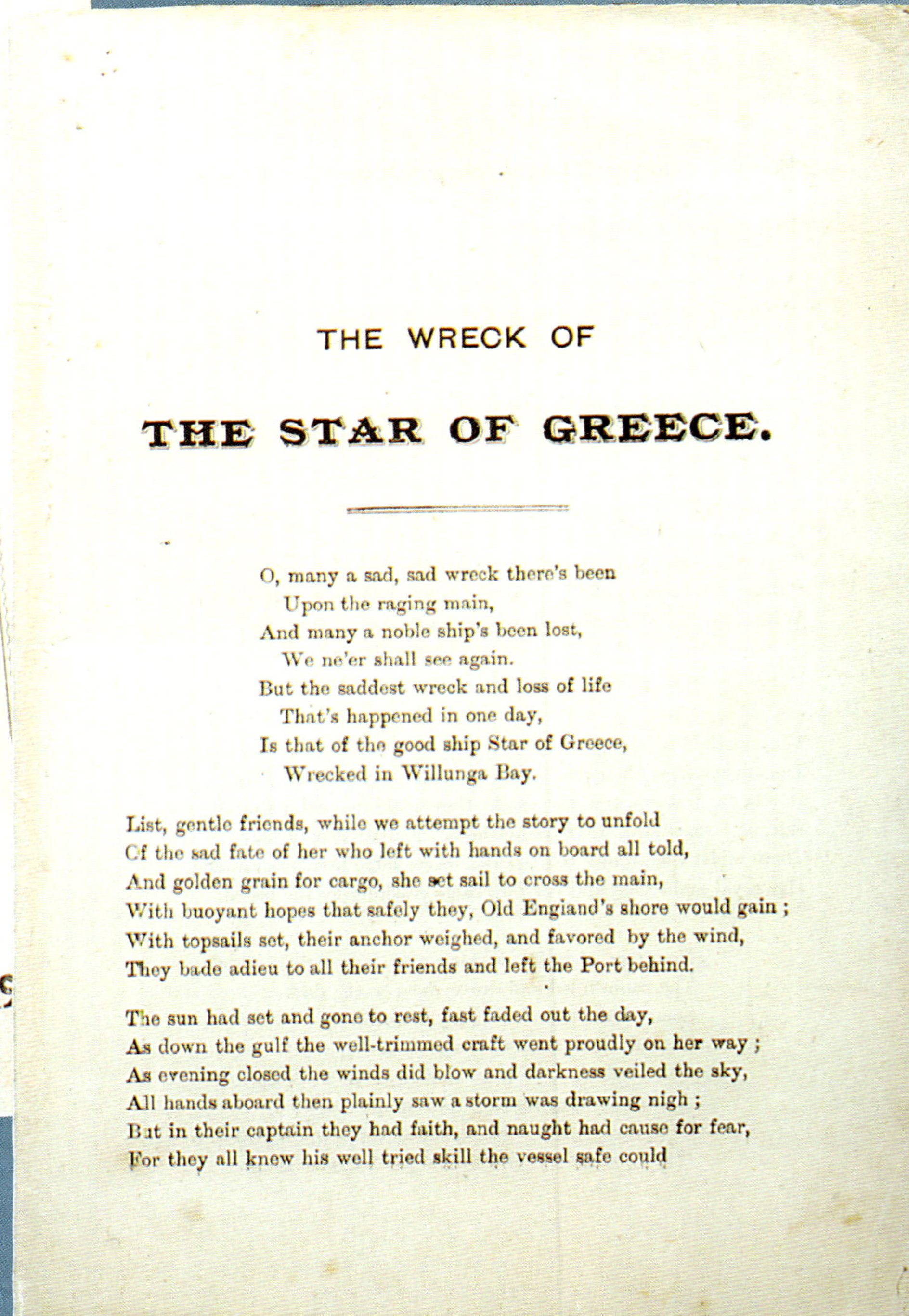

THE WRECK OF

THE STAR OF GREECE.

O, many a sad, sad wreck there's been
Upon the raging main,
And many a noble ship's been lost,
We ne'er shall see again.
But the saddest wreck and loss of life
That's happened in one day,
Is that of the good ship Star of Greece,
Wrecked in Willunga Bay.

List, gentle friends, while we attempt the story to unfold
Of the sad fate of her who left with hands on board all told,
And golden grain for cargo, she set sail to cross the main,
With buoyant hopes that safely they, Old England's shore would gain;
With topsails set, their anchor weighed, and favored by the wind,
They bade adieu to all their friends and left the Port behind.

The sun had set and gone to rest, fast faded out the day,
As down the gulf the well-trimmed craft went proudly on her way;
As evening closed the winds did blow and darkness veiled the sky,
All hands aboard then plainly saw a storm was drawing nigh;
But in their captain they had faith, and naught had cause for fear,
For they all knew his well tried skill the vessel safe could

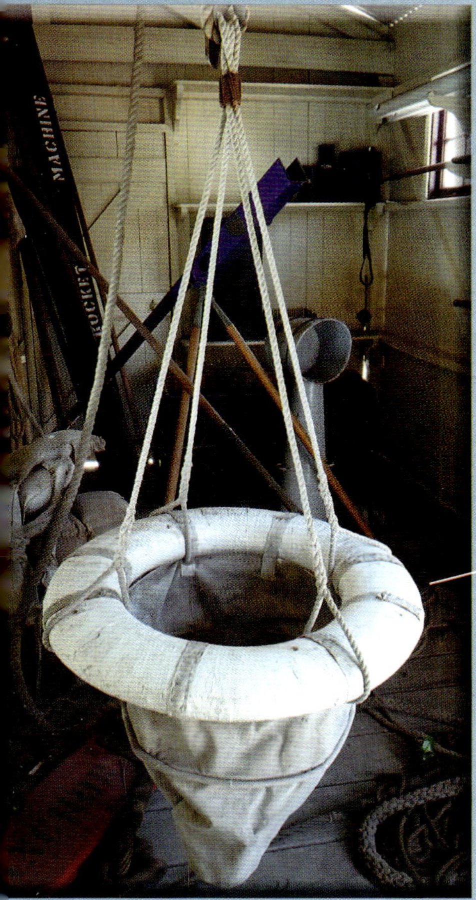

The rocket gear that may have saved the *Star of Greece* crew is demonstrated in these pictures from Flagstaff Hill Maritime Village. A 'rocket' with a line attached is fired from shore over the vessel. Heavier line is attached, and a Breeches Buoy is pulled between the ship and shore with a person within the life ring.

Loch Vennachar

The loss of the clipper ship *Loch Vennachar* in 1905 started a mystery that would remain unsolved for over seventy years.

On 6 September 1905 the steamer *Yongala* came across a ship 260 kilometres west of the Neptune Islands, and after changing course to get closer saw that it was the *Loch Vennachar*.

The vessel appeared in good condition and signalled " Loch Vennachar, please report me all well". She was not seen again.

On 18 September the steamer *Governor Musgrave* left Port Adelaide to search for the now overdue *Loch Vennachar*, but found neither the vessel nor any evidence of wreckage.

On 29 September the ketch *Annie Watt* reported picking up a reel of blue printing paper in Gulf St. Vincent, which was established to be part of the missing vessel's cargo. The previous night wreckage had been reported at Cape St.Albans on the eastern end of Kangaroo Island, and the *Governor Musgrave* once again left to search for the wreck and any survivors.

Extensive wreckage was found along both the eastern and southern sides of Kangaroo Island, but the location of the wreck could not be found. The wreckage was substantial and positively identified as being from the *Loch Vennachar. The Governor Musgrave* eventually abandoned the search and returned to Adelaide on 12 October.

The discovery of a badly decomposed body and huge amounts of heavy wreckage in West Bay on 26 November lead to the new theory that the ill fated vessel had in fact been lost at the western end of the island. Wreckage included major ship fittings, more reels of printing paper and hogsheads of whiskey.

The *Governor Musgrave* was sent to West Bay in early December, salvaging everything of value for the underwriters. The actual location of the wreck remained a mystery, although the amount of new wreckage washing up into West Bay proved that the remains of the vessel were in that general area.

It was not until February 1976 that the final resting place of the *Loch Vennachar* was found, just north of West Bay, by divers from the Society for Underwater Historical Research (SUHR) and two local divers. The wreck lay in 10 metres of water under 30 metre cliffs.

The divers had been searching this area for the wreck when part of a fire brick was found high upon the cliffs. Knowing that part of the *Loch Vennachar's* cargo consisted of bricks, the search was successfully concentrated in this region.

A full underwater archaeological survey of the wreck was conducted by SUHR in February 1977, with support from the Kangaroo Island community, the State Government and many of its Agencies, including the Police Underwater Recovery Squad, and massive support from private enterprise and individuals.

The sheer logistics involved getting a team of over 3 dozen divers over to Kangaroo Island and set up for two weeks at the extreme western end of Flinders Chase National Park were enormous. The expedition had to be completely self sufficient, and this meant taking over two 5 ton vehicles loaded with everything necessary to live and dive for a fortnight. The truck contents included freezers, food, power generators, air compressors, tents, inflatable boats, outboard engines, cliff climbing gear and diving equipment. Even water had to be brought in by tanker.

The outcome of the survey was successful, notwithstanding bad weather, and resulted in a complete record of the wrecksite being produced.

The bower anchor was raised in 1980 and after the necessary treatment to prevent corrosion, was put on display in Flinders Chase.

After many years the mystery of the location of the *Loch Vennachar* has been solved, but the reasons for its loss followed the vessel to its watery grave.

The *Loch Vennachar* was a 3 masted iron ship of 1552 tons, built in Glasgow in 1875, measuring 76 metres in length. At the time of her loss she had a crew of 26 and was under the command of Captain W.F.Hawkins. The *Loch Sloy*, a ship from the same line, was also wrecked on Kangaroo Island in 1899, with the loss of 31 lives. The *Loch Sloy* was an iron barque of 1280 tons, measuring 69 metres in length.

Searchers at West Bay, Kangaroo Island

The postcard safely reached its destination.
Sadly, William Barry did not.

New York
30/5/05

Dear old Pal,

Am present in above [SS Caledonia].
Trust you are in capital health spiritual & otherwise.
Leaving Glasgow in "Loch Vennachar" for Adelaide approximately June 10th.

Always Yours Wm Barry

Crew on an earlier, more fortunate, journey.

Top left Searchers on '*Governor Musgrave*.
Top right Author *left* with John Boon, great grandson of Captain Hawkins.
Bottom The wreck lies below 30m cliffs, constantly pounded by the southern ocean.

Koenig Wilhelm II

The rugged Western Australian coast is not the only location of the wrecks of old Dutch ships.

The discovery of gold in Victoria resulted in massive numbers of people rushing to the goldfields of Bendigo and Ballarat, hopeful of making a quick fortune. Many of the miners came from overseas, including China.

The Victorian Government, in an attempt to increase revenue, or to perhaps slow down the numbers of Chinese immigrants, introduced a poll tax of ten pounds ($20) on all Chinese entering Victoria. This tax was quite a significant amount, being the equivalent of the fare from China to Australia. In these pre Commonwealth days States were each independently responsible for their own immigration policies.

In order to avoid payment to the Victorian Government, many Chinese landed in South Australia, particularly at Robe, and walked overland to the Victorian goldfields, often having been deceived about the distance. Nearly 20,000 Chinese were landed at Robe between 1856 and 1858. Many of the immigrant ships were barely seaworthy, and all were overcrowded.

The *Koenig Wilhelm 11* was one of these vessels, arriving at Robe on 15 June 1857, with 397 Chinese passengers and 26 crew. The pumps had to be worked continuously throughout its voyage, often by the passengers, in order to keep afloat. The conditions aboard the ship were appalling, with several passengers being allocated to each bunk, treated more like cattle than people.

On 30 June a gale struck. The *Koenig Wilhelm 11* had unloaded its passengers, but was still at anchor in Guichen Bay at the time. The vessel was not seaworthy enough to put to sea, particularly without passengers to work the pumps, and risked being sunk while at anchor as the gale worsened, so Captain Geizen decided to beach it on Long Beach, 5 kilometres from Robe.

Signal Cannon recovered from the wreck.

The Captain was still on board the beached vessel when the line holding the lifeboat with the 25 crew snapped, stranding him. As it turned out he was lucky as the lifeboat was swamped, throwing the crew into the wild sea. Only 9 survived. The captain eventually made it to shore on a line that he was able to float ashore on a cask. The ship quickly broke up and became a total loss.

The bodies of the 16 drowned crew were buried in the nearby sandhills in coffins made from the timbers of the wreck. The bodies were reburied 76 years later after the original graves were exposed by the continual shifting of the sand by the wind.

The *Koenig Wilhelm 11* was a Dutch barque of about 800 tons, built in 1840 at Kinderdijk. Some records show the vessel's name as *Koenig Willem 11*. A cannon from the wreck is mounted on display in Robe.

A Joss House, or temple, in Bendigo is one of the few surviving monuments of Chinese immigration from this period in Australia's history. Old broken Chinese relics are still occasionally found by beachcombers at the water's edge, not only near Robe, but in Gulf St. Vincent or Victoria. Such discoveries usually start rumours of ancient Chinese ships sunk off Australia by those not familiar with our history.

QUEENSLAND

Pandora

The mutiny on HMS *Bounty* in 1789 resulted in HMS *Pandora* being sent to the South Pacific with the task of capturing Fletcher Christian and his fellow mutineers.

Having taken control of the *Bounty*, the 25 mutineers cast its master, Captain William Bligh, adrift in a 7 metre longboat with eighteen of the crew and limited supplies. Against all odds, in a remarkable feat of seamanship and navigation, Bligh and all but one of the crew survived, eventually reaching Timor, and from there Batavia then England.

The Admiralty took a serious view of the situation and determined to find and retake the *Bounty*, and to bring the mutineers to justice. It was a clear and unequivocal message that the Navy would not tolerate mutiny and would go to extraordinary lengths to capture wrongdoers.

The 35 metre HMS *Pandora*, under the command of Captain Edward Edwards, was chosen for the task, leaving England on 7 November 1790 with a crew of 132 men. *Pandora* carried an unusually large complement of crew and armaments, sufficient to overpower the *Bounty* once it was found and to crew it for the expected return journey.

Fourteen mutineers were found in Tahiti and captured in April 1791. A number of these men remained loyal to Bligh but had remained on the *Bounty* once the longboat could take no more on board, and were later found innocent at their trials. The next four months were spent searching the many small islands for the *Bounty* and other mutineers.

Christian and the remaining mutineers were not to be found, having sailed to Pitcairn Island, then unchartered, and destroyed any evidence of *Bounty* by burning it. Their descendants, and one surviving mutineer, were found by sealers in 1808 on the American ship *Topaz*.

Those captured were kept in what became known as *Pandora's* Box, a makeshift prison 3.3 m by 5.4 m by 1.5 m high. After months of fruitless searching Edwards left the South Pacific on 15 August for Timor, reaching the Great Barrier Reef eleven days later.

Describing their conditions, prisoner Peter Heywood stated:

"We were all put in close confinement, with both legs and hands in irons, and were treated with great rigour, not being allowed ever to get out of this den. And, being obliged to eat, drink, sleep and obey the calls of nature here, you may form some idea of the disagreeable situation I was in."

On 28 August, while attempting to navigate through what was believed to be a safe passage, *Pandora* struck a submerged outcrop of reef at low tide, and quickly began taking water.

The depth of water around the reef was 30 metres, leading Captain Edwards to believe that the ship was in no danger, but the outcrop of reef itself was only 5 metres below the water at high tide, and significantly less at low tide.

Heavy guns were dumped overboard to lighten the ship and the crew worked the pumps while attempts were made to repair the hull. The failure of one of the pumps next morning doomed their efforts and the ship quickly sank.

Pandora's boats had been launched and provisioned, and stood ready to pick up those on the sinking ship. Three prisoners had been let out of the box to help with the pumps, and as the ship started to sink, William Moulter, the bosun's mate, unlocked the hatch to give the remaining prisoners a chance to survive. Four could not get free of their manacles and drowned, together with thirty one of *Pandora's* crew.

The survivors left for Timor in four boats, having found and rescued *Pandora's* cat from a mast protruding above the water on their last visit to the wrecksite. After an uneventful trip they reached Timor on 13 September. One of the survivors, Lieutenant Thomas Hayward, had been one of the original *Bounty* crew who had also survived the original longboat journey with Bligh in 1789.

Silver pocket watch believed to belong to *Pandora's* surgeon, George Hamilton.

Yongala

122 people, 49 passengers and 73 crew, lost their lives when the steam ship SS *Yongala* disappeared during a cyclone in 1911, south of Townsville off Cape Bowling Green.

The ship was travelling to Cairns, having left Melbourne on 14 March, with Captain William Knight stopping at Brisbane and Mackay to transfer passengers and cargo on and off.

At Mackay 50 tons of cargo was discharged and new cargo taken on board, including a race horse called Moonshine, which was to be delivered to Townsville.

Leaving Mackay on 23 March, *Yongala* was sighted by the Dent Island lighthouse keeper in the Whitsunday Passage, in what was to be the last time the ship was seen. The weather was quickly worsening, but *Yongala* carried no radio and could not be contacted. Ironically a wireless had already left England for installation in the vessel.

The 107 metre steel ship was in excellent condition and with an experienced master aboard was regarded as capable of riding out a cyclone, so its late arrival did not cause undue concern initially. By 26 March *Yongala* was posted as missing and full searches were gotten underway by the Queensland Government.

Wreckage started being washed up, but no trace was found of the ship or the people on board. The body of Moonshine was, however, found at the mouth of Gordon Creek, near Townsville.

A reward was put up for information leading to the discovery of the lost ship, but it was not found and the reward withdrawn. A *Yongala* Distress Fund was also established to help the families of those lost.

An Inquiry into the loss, in the absence of survivors or witnesses, focussed its attention on the condition of the ship and the experience of its Captain. The Inquiry established the seaworthiness of the vessel and also the ability and careful character of Captain Knight.

In 1947 the HMAS *Lachlan* surveyed an obstruction eighteen kilometres off Cape Bowling Green reported by a minesweeper in 1943. Using an echo sounder and submarine detection instruments *Lachlan* identified the object as a large wreck about 100 metres in length, but took no further action to investigate it further.

It was 1958 before two divers found the wreck, resting in 30 metres of water. A safe found on the wreck was brought to the surface, and although empty, its serial number confirmed that it came from the *Yongala*.

Many relics from the wreck, including the bell, are on display in the Townsville Maritime museum. The wreck itself is a major tourist attraction, with over then thousand divers visiting it each year.

In 1905, *Yongala* was the last ship to sight the *Loch Vennachar* in South Australian waters before it disappeared in mysterious circumstances.

"S. S. Yongala"

1903

Captain Rees of the *Yongala* received a signal from Captain Hawkins on 6 September 1905, 260 kms west of the Neptune Islands "*Loch Vennachar* please report me all well." It was to be the last sighting.

Gothenburg

In 1875 Palmerston, now called Darwin, was a major city, but relied upon the southern states for its business and banking. The previous year *Gothenburg's* owners entered into a 2 year contract to provide 10 trips between Darwin and Adelaide to satisfy the growing commercial needs of the region.

An iron steamship, *Gothenburg* was also rigged with sail, and regarded as a particularly fast vessel. Leaving Darwin late February 1875, Captain James Pearce carried 3,000 ounces of gold from the Timber Creek mines, as well as about 100 passengers and 34 crew. Some reports show 84 passengers, others 92.

Those aboard included the French vice Consul, prisoners, judges and miners.

On 24 February *Gothenburg* was heading south along the eastern coast of Queensland in deteriorating conditions, with a cyclone building. With restricted visibility Captain Pearce steered away from the coast, but ran onto Old Reef south of Townsville while travelling at full speed.

At first it was expected that *Gothenburg* would float off as the tide rose, but the boiler fires were extinguished by the sea coming in through the damaged hull. Without power and with the conditions worsening the ship heeled over onto its side. Many were lost overboard.

Attempts to launch lifeboats were made and many more passengers and crew swept away in the process. A few people succeeded in getting into boats, but it was impossible to carry out a proper evacuation in the conditions that prevailed and with the vessel breaking up.

102 lives were lost including the Captain. Some reports show the number of lives lost to be 112. The few survivors were rescued from nearby islands by search vessels from Townsville.

Captain Pearce

About 20% of the gold being carried on the *Gothenburg* was later salvaged by salvage operations conducted from the steamers *Bunyip* and *Porpoise*.

Captain Pearce's son, Tom, was one of only two survivors of another significant wreck, the *Loch Ard*, off the Victorian coast in 1878.

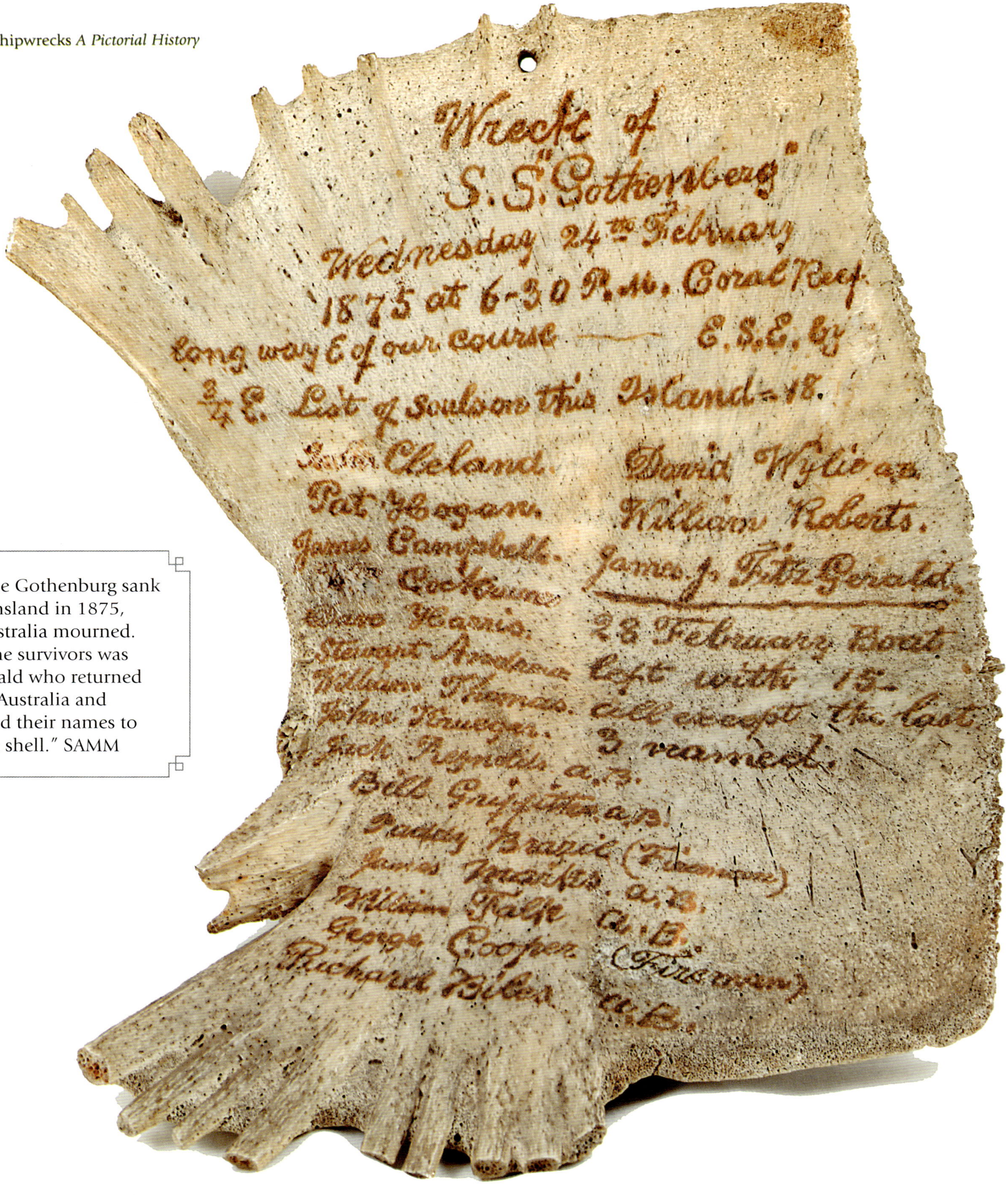

"When the Gothenburg sank off Queensland in 1875, South Australia mourned. Among the survivors was J.J Fitzgerald who returned to South Australia and committed their names to this turtle shell." SAMM

Gothenburg was carrying several dignitaries, including the French Vice-Consul, the Hon. T. Reynolds, former Premier of South Australia, Mr. R. Wells, editor of the Northern Territory Times and judges.

Cherry Venture & Stirling Castle

The significance of the wreck of the *Cherry Venture* results from the fact that it is high and dry on the Sunshine Coast, north of Noosa, and seen by many thousands of people annually.

On its way to New Zealand from Brisbane, the 1600 ton Singapore registered *Cherry Venture* was caught in the winter gales of 1973, with swells recorded at over fifteen metres. With no cargo aboard, the lightly loaded vessel could make little headway to get away from shore. Even under full steam the *Cherry Venture* was washed towards shore, with its propeller being lifted out of the water by the size of the waves.

Pushed in towards shore on the morning of 8 July, the Captain sent a May Day call. The Air Force rescue helicopter battled 120 kph winds, but could not get to the ship. Two anchors laid by the Captain did not hold the vessel, with both cables snapped by the force of the sea.

After landing at Maroochydore, the helicopters were guided north towards the ship by following gullies to avoid the worst of the gale.

At 1.00pm the *Cherry Venture* was washed up onto the beach, with all 24 crew safe, although badly shaken. Even if the helicopters had got to the vessel it is highly unlikely that they could have removed the crew while at sea because of the force of the gale.

Salvage attempts were unsuccessful and the *Cherry Venture* is now a popular tourist site for the many four wheel drivers en route to Fraser Island.

As the condition of the ship has deteriorated visitors can now observe the ship but can no longer climb up onto it.

Fraser Island itself was named after the survivor of a shipwreck. In 1836 the *Stirling Castle* was lost in a storm and the seventeen survivors took to 2 boats and sailed towards Moreton Bay to the south.

Eliza Fraser was heavily pregnant and went into labour during the days following the wreck while bailing water from the small boat. The baby was born alive but died very soon after. After about 3 weeks they landed on what was then known as Great Sandy Island.

Although many Aboriginal tribes were friendly, the one on the island was not and took Eliza Fraser and others captive, eventually killing her husband. Some of the survivors were, over time, sent to other tribes, including on the mainland.

Three months later friendly Aborigines met a convict, John Graham, at Moreton Bay and told him of the "white ghosts" as they were described, living on Great Sandy Island. Prison officials were advised and a group including John Graham went north, finally rescuing Eliza Fraser. Graham, who was nearing the end of his sentence, was released early for his rescue efforts.

Eliza Fraser returned to England where she married the Captain of the ship *Mediterranean Packet*, the ship that carried her back home.

CHERRY VENTURE

LCANNIA

River Murray

Riverboats faced many and varied perils, quite different to those of ocean going vessels.

In times of drought boats could be stranded for many months if they were unlucky enough to be caught away from a deep part of the River. Commercially this was disastrous.

At the other extreme floods brought their own danger. Fast flowing waters made navigation difficult, particularly for the boats with lesser powered engines. The changing shape of the River during times of flood also raised the significant risk of boats hitting snags such as fallen logs, increasing the ever present risk of sinking. Rapidly receding flood waters could also leave the boat stranded.

Snags were a constant a hazard to the riverboats, whose captains relied upon experience and hand drawn charts. Skippers of each boat were valued not just because of their ability to handle a vessel but also because of their navigation experience and knowledge of the River hazards. Many, if not most, of the boats on the River had their hulls punctured by unexpected underwater snags, although fortunately most were able to be repaired or recovered.

Wood fired steam engines also created the ever present risk of fire aboard boats. Many vessels caught fire with several being total losses. The quickest and most effective way of quickly extinguishing a blaze was often to sink the vessel, enabling the later recovery of the remains and hopefully its cargo.

One of the rare cases involving the loss of life was the Paddle Steamer *Bunyip*.

PS *Bunyip*, while towing two barges, caught fire about 12 kilometres above Chowilla Station in 1866.

The fire rapidly took hold and all attempts to put it out were unsuccessful. Captain Randell decided to run the vessel into the bank in an attempt to save it and the 20 persons on board.

Unfortunately the steering jammed and the vessel circled helplessly in the middle of the river for sometime before eventually drifting on to one of the river banks when its engines stopped.

One woman passenger and her child died in the flames and two crew members also drowned while attempting to swim ashore. Most passengers and crew suffered burns.

The *Bunyip* was a total loss as were the barges it was towing. Captain Randell later managed to salvage one of the hulls and rebuilt it into the sailing vessel *Waterlilly*, which was also used on the River and lakes before itself sinking in Lake Alexandrina in 1903.

The development of rail and road transportation saw the demise of the commercial riverboat trade, and many boats and barges were simply abandoned when they were no longer commercially viable. Many of their remains can now be found along the banks the length of the River Murray, particularly in the vicinity of old riverports.

Wartime wrecks

HMAS *Sydney II*

The discovery of the wreck of the HMAS *Sydney* in 2008, sixty seven years after its loss with all hands, marked an end of an era of uncertainty.

The battle between the Australian warship *Sydney* and the German raider HSK *Kormoran* off Western Australia on 19 November 1941 ended with both ships sinking, and started decades of controversy about the events of that day.

The loss of the *Sydney* with its superior size and firepower to a lesser raider could not be accepted by many. Wartime Government censorship of media coverage of the loss compounded rumours, with even families of those lost only receiving bereavement letters which did not provide details.

With all of Sydney's 645 crew lost the only accounts of the tragedy came from German survivors, with many believing that the lack of Australian survivors could only have been through German lifeboats not rescuing them, or worse.

Searches for both ships were fruitless until 2008 when the *Finding Sydney Foundation* secured the funds and expertise to mount what proved to be a successful expedition.

Captain Theodor Detmers on *Kormoran* was 220 kilometres south west off Carnarvon when at about 5.00pm his lookout reported smoke on the horizon 25 kilometres away.

Kormoran's role was to attack merchant shipping in order to force Australian warships away from the main battle zone to defend their ships. Believing the smoke to be from several vessels accompanied by an escort, Captain Detmers turned away.

Sydney, returning to Fremantle from Sunda Strait had, however, sighted *Kormoran* and by 6.00pm was within one kilometre of the ship which was disguised as a Dutch merchant vessel, the *Straat Malakka*.

Unable to provide the Allies secret call sign requested by *Sydney's* Captain Joseph Burnett, Captain Detmers changed the Dutch flag for the German ensign and immediately opened fire, destroying

Sydney's bridge, gun director tower and forward turrets with the first three salvos.

Sydney's initial shells were high, but the rear gun turrets hit *Kormoran's* engine room, causing uncontainable fires which rapidly spread through the ship.

Sydney was also ablaze, crippled by a torpedo hit to the bow, and slowly steamed away to the south east, firing four last torpedos as she passed near the stern of *Kormoran*. None hit.

Kormoran was also burning, with fires out of control, when at 7.30 pm Captain Detmers gave the order to take to the lifeboats. By this time *Sydney* was over 5 kilometres away, still ablaze and moving to the south east. *Kormoran's* crew saw no lifeboats launched from Sydney.

There were no survivors from Sydney and very little wreckage. On 6 February 1942 a damaged life raft known as Carley float containing a body washed up on Christmas Island 2,500 kilometres from the battle site. Authorities now acknowledge the raft as being from *Sydney*, although this was not so at the time of its discovery.

Kormoran's 317 survivors of the 390 crew were rescued on 27 November by the merchant ship *Centaur*, which took lifeboats in tow, being unprepared to take them on board as her crew were not armed.

It was to be March 2008 before either vessel was seen again. Leaving Geraldton on 3 March the SV *Geosounder*, hired by the *Finding Sydney Foundation*, located first the *Kormoran*, then a day later on 17 March, *Sydney*, 22 kilometres away on the seabed at a depth of almost 2,500 metres.

Specialist technical analysis of the underwater remains of both ships has verified the accounts of German survivors, making it clear that catastrophic damage through explosion and fire resulted in *Sydney* being unable to launch lifeboats as it limped towards the far distant Western Australian coast.

The extraordinary work undertaken in searching for the wrecks and analysing their remains pays tribute to those who lost their lives, and allows their families and loved ones a measure of closure after decades of uncertainty.

Kormoran survivors

One of *Kormoran's* guns

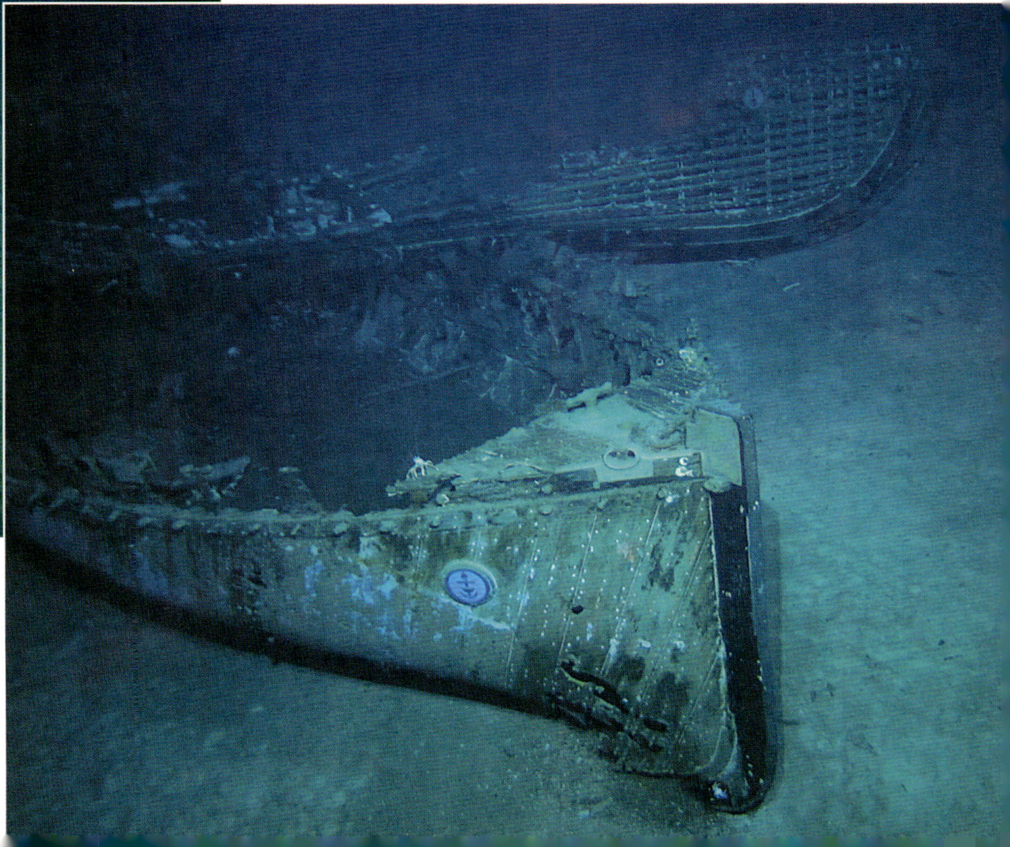

One of *Sydney's* lifeboats

Sydney's crew

HMAS *Perth*

Adelaide diver David Burchell's discovery of the HMAS *Perth* in Sunda Strait is an extraordinary achievement.

Using his own resources and with only basic scuba equipment, he was assisted by local Indonesian fishermen to visit spots known to them. At the extreme limit of his equipment, in heavy currents and diving with limited visibility he found HMAS *Perth* in 1967.

Part of an Allied fleet in the Battle of the Java Sea, *Perth*, under the command of Captain Hector Waller, joined several other warships on 27 February 1942 to block a Japanese convoy. Ordered to escape through Sunda Strait after the loss of several other vessels, *Perth* was making for the south coast of Java when on 1 March Japanese torpedoes hit and sank her, with the loss of 353 lives.

Survivors became prisoners of war, with over one hundred of the crew dying in captivity.

Accompanied by USS *Houston, Perth* had received incorrect intelligence that there were no Japanese vessels in the Strait. The Japanese fleet in fact was ahead of both ships, blocking any escape.

The destroyer *Fubuki*, one of 12 Japanese ships in the vicinity, sighted both allied warships at 10.30pm and shadowed them for half an hour. Soon after 11.00pm *Perth* sighted the Japanese *Harukaze*, then *Fubuki*, and opened fire missing both.

A further six Japanese ships quickly arrived at the scene, with another two called in. A heavy gunfire and torpedo battle was soon underway, with *Perth* striking both *Harukaze* and *Shirayuki*. Just before midnight Perth was hit by fire from *Harukaze* but suffered limited damage.

Running low on ammunition, *Perth* was struck by 4 torpedos from *Shirayuki*, *Murakumo*, *Harukaze* and *Hatakaze* in the space of 15 minutes and sank at 25 minutes past midnight on 1 March. USS *Houston* sank 20 minutes later, having become the full target of all Japanese torpedoes after *Perth* sank.

It was to be 25 years later that underwater explorer David Burchell determined that both *Perth* and *Houston* could be within reach of a diver using the relatively new SCUBA equipment. The idea of finding *Perth* grew out of Japanese proposals to search for and recover ships sunk during the war for their scrap metal value, and the realization that they could rest within diveable reach.

A chance meeting with Gordon Reid, the South Australian President of the Ex-*Perth* Association resulted in discussions with, and moral approval from, the Association for David Burchell to attempt the search and dive.

Although assisted by the Indonesian Government and Navy, the task ahead was enormous and full of unexpected obstacles. The *Aries*, provided as a dive platform for the search and all operations, was a derelict, and even when its engine was repaired, it carried little equipment. After several false leads, it was eventually local fishermen who identified known wreck locations. The personal support of individuals at all levels in Indonesia was a key element to the success of the project.

One of the fishermen's locations proved to be *Perth's* final resting place, with the wreck found while diving in a 5 knot current, at a depth almost double the 40 metre recognised safe limit of SCUBA equipment.

In all thirty dives were made, recovering twenty four relics, including shell cases and navigational instruments, most of which are on display in the Australian War Memorial. Some relics are also in the city of Perth or donated to *Perth* historical and survivor Associations.

David Burchell hands over the voice tube mouthpiece from the old HMAS *Perth* to Captain David Leach, commander of the new HMAS *Perth*

Submarine attack on Sydney Harbour

May 1942 saw the unthinkable happen, with Japanese submarines entering Sydney Harbour and attacking shipping.

As a major base for allied ships and forces, Sydney hosted vessels from the United States, United Kingdom and the Netherlands, as well as Australian naval and merchant ships. The fear of attack resulted in nets being strung across the entrance to the Harbour, which were parted to allow for movement of vessels as necessary.

On 31 May three midget Japanese submarines, each armed with two torpedoes and carrying two crew, were launched at sea from larger submarines, *I 22*, *I 24* and *I 27*, eleven kilometres from Sydney Heads.

An aircraft launched the previous night from one of five large submarines, *I 21*, over fifty kilometres from Sydney had determined the presence of about thirty war ships within the Harbour, warranting an attack. The aircraft was seen by Navy personnel but was not associated with submarines at sea outside the Harbour. The aircraft sank in the rough sea conditions when it returned to I 21, but the aircrew were able to board the submarine and report their sightings.

At 8.00pm *Midget 14* was caught in anti torpedo nets as it attempted to enter the Harbour and was blown up by its crew, rather than be captured.

Midget A successfully negotiated the nets and entered Sydney Harbour, but was spotted and fired upon by the American cruiser USS *Chicago*. The submarine fired both torpedoes, with the first missing *Chicago* and running onto Garden Island, where it failed to explode.

The second torpedo missed the Dutch submarine *K 9*, but exploded under a converted Harbour passenger ferry, the HMAS *Kuttabul*, killing twenty one of those aboard. Two of the dead were Royal Navy, the other nineteen Royal Australian Navy. The ferry was being used by the Australian Navy as a barracks ship and was tied to the wharf at the time.

The third submarine, *Midget 21*, was sighted by HMAS *Yandra* as it neared the nets and fled after six depth charges were dropped. It returned four hours later and was again sighted, resulting in it being attacked with more depth charges. *Midget 14* was recovered the next day and found to have both torpedoes stuck in the firing tubes. Both crew had shot themselves.

The bodies of the four Japanese submariners were given a military funeral and their ashes returned to Japan after the war.

The remains of *Midget 14* and *Midget 21* were assembled to form a display of a single submarine at the Australian War Memorial.

Midget A was not found at the time, but its remains were discovered outside of the Harbour by divers in 2006. It is now also on display in the Australian War Memorial.

Top right sunken *Kuttabul*

Top left Lord Mountbatten visits the War Memorial.

Burial service for naval men killed by submarine attackers

AHS *Centaur*

The sinking off North Stradbroke Island of the Australian Hospital Ship *Centaur* by a Japanese submarine on 14 May 1943 created outrage within Australia. There were only 64 survivors of the sinking, with 268 lives lost, including 18 doctors and 11 nurses.

Operating as a merchant vessel for many years, *Centaur* had been converted to a Hospital Ship in early 1943, and carried clear markings to show its status. Protests were made at the time and proposals initiated to pursue war crimes charges against the submarine's Captain, although it was to be thirty six years before the identity of the submarine involved, I 177, was established.

The conversion to a Hospital Ship completed in the first two months of 1943 and *Centaur* was commissioned in its new role on 1 March. Formal notification of this, including its markings, was made to both the International Committee of the Red Cross and to the Japanese Government.

On its initial voyage as a Hospital Ship *Centaur* carried wounded troops, and wounded Japanese prisoners of war, from New Guinea to Sydney. Resupply occurred in Darling Harbour with *Centaur* leaving to again travel north to Cairns and New Guinea on 12 May, with a crew of 74 and a complement of medical, nursing and field ambulance soldiers from the non combatant Unit 2/12.

Prior to its departure the Captain was alerted to the ambulance soldiers attempting to bring their rifles aboard, creating concern that the ship could be regarded as transporting military supplies. *Centaur's* Captain was eventually convinced that the Hague Convention allowed such rifles to be carried under the provision which provided for defence of the wounded. A search of the cargo prior to departure found no other weapons.

At 4.10am on 14 May *Centaur* was hit by a single torpedo below the waterline blowing a hole into the hull nearly 10 metres across, and setting the ship afire. Most aboard were asleep at the time, and *Centaur* sank within three minutes.

No lifeboats could be launched, although two drifted to the surface as the vessel sunk. *Centaur* was estimated to be 44 kilometres east-north east of North Stradbroke Island at the time, based on its last plotted location at 4.00am.

It was to be 36 hours before the survivors, clinging to the damaged lifeboats and other floating wreckage, were rescued, picked up by the American Destroyer USS *Mugford*, which had been alerted when a RAAF Avro Anson air crew spotted survivors in the water. Sister Ellen Savage was the only woman amongst the survivors.

Mugford continued to search for other survivors until dark before returning to Brisbane, with further ships taking up the search after its departure. No other survivors were located and the search was called off on 21 May.

Francis Martin, *Centaur's* cook, sighted the submarine on the surface while he was floating upon wreckage, and from his description it was identified as a Japanese KD7 Kaidai class.

It was 1979 before the submarine responsible for the sinking was identified as I 177 with the publication of the official Japanese War History. Submarine I 177 had been under the command of Lieutenant Commander Hajime Nakagawa at the time. He never spoke of the sinking of the *Centaur*, but was imprisoned for four years after being convicted for shooting survivors of three British merchant ships in 1944 while in command of another submarine.

The *Centaur* had in 1941 rescued the German survivors of the HSK *Kormoran* after its fateful battle with HMAS *Sydney*, which resulted in the sinking of both ships.

Sister Ellen Savage, the only nurse to survive the *Centaur* sinking.

Bombing of Darwin

The air raids on Darwin have been referred to as Australia's Pearl Harbour, with more bombs being dropped on it than its American counterpart.

The 19 February 1942 bombing runs were the first of sixty four such raids, causing significant damage to the city and sinking ships in the Harbour and at nearby Bathurst Island. Twenty five planes were lost, most being destroyed on the ground.

The attacking planes were launched from four aircraft carriers of the Imperial Japanese Navy, *Akagi*, *Hiryu*, *Kaga* and *Soryu*, with 188 planes in the first wave, and 56 in the second that day.

The first wave lasted just over 40 minutes and the second a mere 25 minutes.

Darwin had minimal defence capability with few guns and fewer planes, and caught unexpectedly, suffered enormous shipping losses. A message half an hour before the air strikes that a large fleet of aircraft was in the vicinity was received from Bathurst Island, but the advice was presumed to relate to 12 US planes that had just left Darwin.

243 persons, military and civilian, were killed on that first day and over 400 injured.

Most of the forty five ships in the harbour suffered damage and eight were sunk. Ships lost were the troop transports MV *Neptuna* and SS *Zealandia*, the patrol boat HMAS *Mavie* and merchant vessels SS *Mauna Loa* and *Kelat*, the UK refueller *British Motorist*, and the American destroyer USS *Peary* and transport USAT *Meiggs*.

The Australian Hospital Ship *Manunda* suffered damage, with 12 of her complement killed and 47 injured. Despite the damage and losses suffered, *Manunda's* crew rescued others from sunken ships and continued to treat the wounded.

Four Japanese aircraft are believed to have been brought down by anti aircraft fire.

At Bathurst Island the merchant ships *Florence D* and *Don Isidro* were sunk, and a USN Catalina was shot down. The *Florence D* was located in 2006.

The Navy base at Darwin was for all intensive purposes abandoned after the 19 February raids, with most ships moved to other ports, while defence capability focussed upon an improved air force presence.

Japanese raids continued on Darwin and other cities and towns of northern Australia until late 1943, and while the feared Japanese invasion never happened, significant military resources were tied up, affected the overall war effort.

REPORT ON (First) AIR RAID - DARWIN 19 Feb 42.

At 0959K 89 Japanese Aircraft approached from the North and high level bombed, dive bombed and machine-gun strafed Darwin. There was no warning of approach and planes were almost over the area before they were recognised as hostile aircraft. First warning of hostilities was dog-fights in progress over Fanny Bay. Planes came over in various formations of Heavy Bombers, Dive Bombers and Fighters.

---------oOo---------

Main targets were:-
Convoy in the Harbour.
R.A.A.F. 'Drome.
Harbour Installations.
Oil Tanks.

Interception was made by 2 P40E's, which at the time were carrying out high level patrol. 7P40E's took off to intercept.

H.E. and Fragmentation type bombs were dropped - ratio 3 H.E. to 1 "F" type.

---------oOo---------

Damage.

Extensive damage was done to R.A.A.F. hangers and service installations. Post Office and surrounding buildings destroyed and Kahlin Hospital damaged by 1,000lb bombs.

Ships Sunk.

U.S. Destroyer	"Peary"
Merchant Ships	"Mavie"
	"Zealandia"
	"British Motorist"
	"Mauna Loa"
	"Neptuna"
	"Meigs"
Ships Damaged	"Portmar"
	"Barossa"
	"Manunda"

The "Neptuna", which was loaded with depth charges and ammunition blew up at the wharf, causing considerable damage.

Associated with the attack on Darwin township, the following additional losses were sustained at sea:-

"Don Isidro" was attacked by enemy bombers in the Arafura

First offical report of bombing.
(Aust War Memorial)

Norfolk Island

HMS *Sirius*

HMS *Sirius* carried Captain Arthur Phillip to the new convict colony of New South Wales, reaching Botany Bay in January 1788 after a journey of eight months. The flagship of the eleven vessel First Fleet, *Sirius* was a 20 gun ship, originally built in London in 1780 as the East Indies trading vessel *Berwick*.

One month later the Union Jack was hoisted on Norfolk Island, 1600 kilometres off the eastern coast of Australia, proclaiming it a British Colony. The colony consisted of twenty two people, including fifteen convicts, and was established in the area now known as Kingston.

The tall and impressive pines of the island were ideal for ship's masts and highly valued, the flax plants were commercially important for sail making and its hemp provided an alternative supply for rope making, necessary after Russia, Britain's main supplier, severely limited its availability.

Norfolk Island, first recorded by Captain Cook in 1774 while on HMS *Resolution*, is only approximately 8 kilometres long and five kilometres wide. It had been previously visited by Eastern Polynesian sailors about three centuries earlier, but no trace of them existed by Cook's visit.

Nine transport ships left New South Wales in 1788 after unloading their human cargo, leaving the new colony dependent upon HMS *Sirius* and *Supply* for contact with the world beyond their shores.

Governor Arthur Phillip's instructions included the requirement to colonise Norfolk Island, and not just New South Wales, in order to prevent foreign access to the island's valuable natural resources.

Following its settlement more convicts were located there, and the island became a farm providing supplies for the colony at Botany Bay. A major problem, however, was the lack of a safe harbour, and a coastline that was almost impossible to access from the sea due to its high cliffs.

Failing crops in New South Wales and a desperate food shortage resulted in a decision to relocate some of the population to Norfolk Island aboard the *Sirius* in 1790.

Leaving New South Wales on 6 March 1790 under the command of Captain John Hunter, *Sirius* arrived at Norfolk Island a week later, only to be wrecked on the reefs of Emily Bay in a gale. No lives were lost and much of the cargo and ship's fittings were salvaged.

In 1856 nearly two hundred descendants of the *Bounty* mutineers also moved to live on Norfolk Island as they outgrew Pitcairn Island.

Navigational Instruments

Early navigation simply involved staying in sight of the coastline, but clearly this was extremely limiting.

The development of the magnetic compass provided rudimentary ability to venture longer distances out of sight of land. The dry land compass was refined to a liquid dampened marine variation, with the liquid having the effect of stopping the otherwise wild swinging of the compass on a moving deck. While this assisted with direction, it did nothing by itself to identify an actual position.

Cartography, or drawing of maps, allowed details of journeys to be recorded and provided for the documentation of significant coastlines, islands, reefs and other objects of interest.

As journeys extended further the charts being drawn included locations based on the direction travelled, with the distance measured by an estimate of speed and time.

The development of a means to identify latitude was a major advancement, with it being possible to estimate distance north or south of the equator by measuring the angle between the sun and the horizon at noon. The accuracy of this measurement was greatly affected by clouds obscuring the sun, and the heaving of the deck from which the measurement was being taken during anything but calm conditions.

The sextant and its assorted predecessors such as the astrolabe and quadrant were developed to measure angles, principally between the planet's surface and celestial objects, such as the sun or North Star. The sextant could also be used to measure angles between flat objects such as features on a coastline, enabling their placement accurately on charts.

While latitude could be measured with some degree of confidence, longitude, or distance east or west, was another matter. The development of the marine chronometer finally enabled this to be calculated. The chronometer is a very accurate time piece, which when used with tables and celestial sightings could allow a calculation of east – west distance. Its development took John Harrison over 30 years.

The ship's log was also developed to allow speed to be measured through the water, enabling measurement of approximate distance travelled over a period of time. By releasing a floating object on the end of a knotted line into the water, it was possible to counts the number of knots that went past an observation point during a known period of time, thus calculating the ship's speed. The nautical term knot as a measure of speed developed from this.

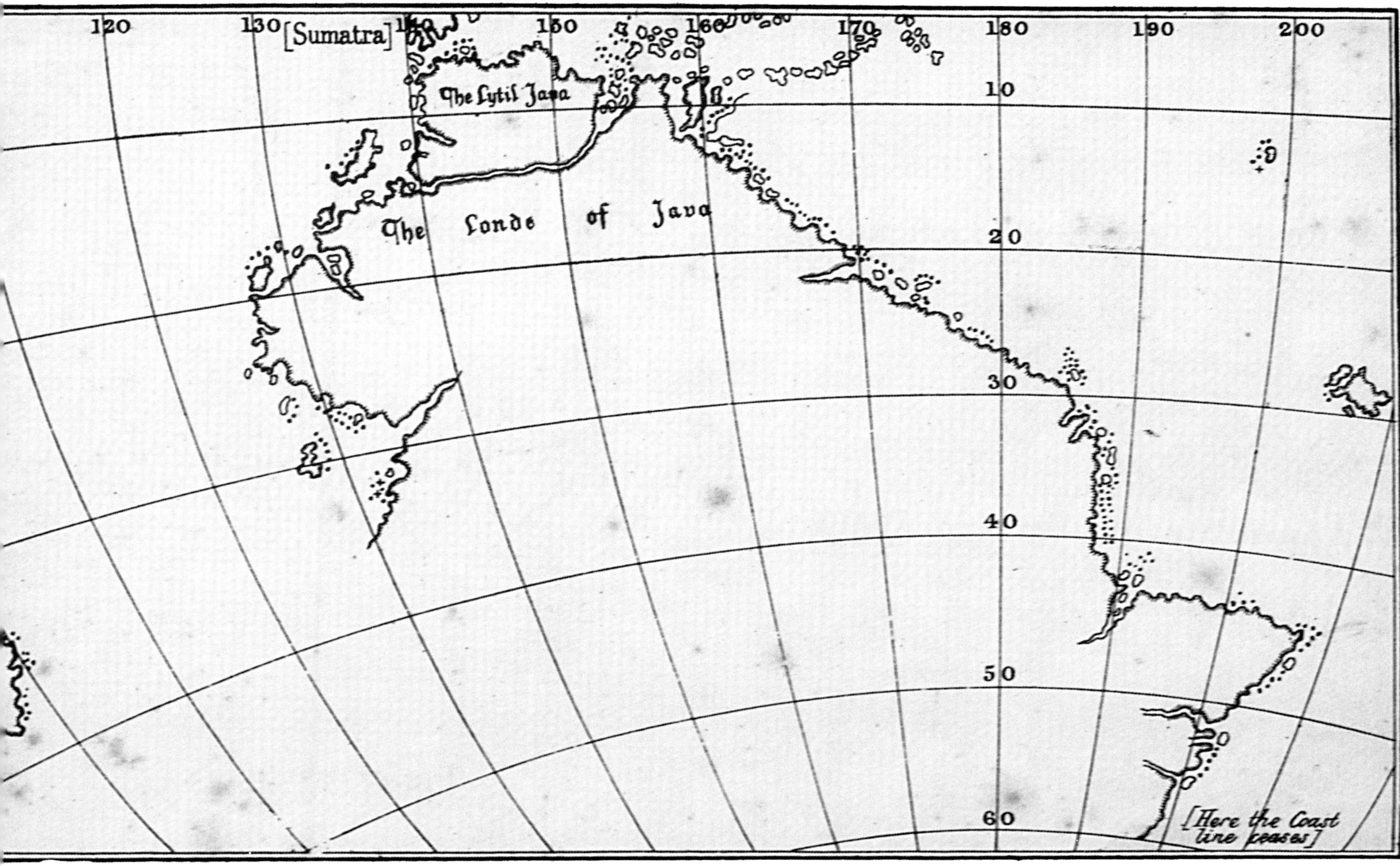

Maps were vital navigational aids. The Rotz Atlas published in1536 shows the Australian coastline as part of Java (now Indonesia.) The reason why the Australian coastline makes a sharp 90º bend eastwards on this map is due to the fact this part of the map was incorrectly joined to the previous section.

Three Eras of Maritime History

Robe's Royal Circuit is a unique display of different elements of our maritime heritage. It reflects the navigation of Flinders and Baudin, a cannon from a wreck which bought Chinese emigrants to the Victorian goldfields through South Australia and wartime mines.

NATIONAL TRUST
THIS GERMAN MINE WAS FOUND ON
THE BEACH ABOUT 10km SOUTH
OF CAPE JAFFA LIGHTHOUSE
ON 18TH JULY 1941
BY WILLIAM JOHN FLETCHER.
FURTHER HISTORY IS RECORDED
IN THE CUSTOMS HOUSE MUSEUM.

SIGNAL CANNON
FROM
KOENIG WILLEM II
DUTCH BARQUE OF 800 TONS
WRECKED IN A GALE WITH THE LOSS OF 16 CREW
ON LONG BEACH ON 30TH JUNE 1857
AFTER UNLOADING NEAR THIS SPOT CHINESE
PASSENGERS FROM HONG KONG
PLAQUE DONATED BY N. A. GIBBONS

Bibliography

Ball. D "Amazing Australian Shipwrecks" (Angus & Robertson 1987)

Bateson. C. "Australian Shipwrecks Vol.#1" to Vol.# 4

Brown. A. "Ill-starred Captains. Flinders and Baudin". (Crawford House Publishing, 2000)

Burchell, D. "The Bells of Sunda Strait" (Rigby, 1971) (*Perth*)

Chapman. G. "Kangaroo Island Shipwrecks". (*Loch Vennachar*)

Charlwood.D. "Wrecks & Reputations". (Angus and Robertson, 1977). (*Loch Ard & Schomberg*)

Christopher. P " Shores of Tragedy. (Shipwrecks of South Australia)". (*Star of Greece & Clan Ranald*)

Christopher. P. "... to save the lives of strangers. (Famous Shipwrecks of South Australia)". (*Loch Vennachar, Admella, & Koenig Wilhelm 11*)

Christopher. P. "South Australian Shipwrecks. A Data Base 1802 – 1989".

Christopher. P. "Australian Riverboats. A pictorial history". (Axiom 2006) (*River Murray*)

Christopher. P. "Paddlesteamers and Riverboats of the River Murray". (Axiom 2000)

Edwards. H. "Islands of Angry Ghosts". (Angus & Robertson. 1973). (*Batavia*)

Fitzgerald. L. " Java La Grande. The Portuguese discovery of Australia". (The Publishers Pty Ltd, 1984)

Gesner. P. "Pandora" (Queensland Museum, 1991)

Green. J. "Treasures from the *Vergulde Draeck* " (WA Museum, 1974).

Halls. C. "Australia's Worst Shipwrecks". (*Star of Greece*)

Heinrich. R. "Wide Sails And Wheat Stacks".

Henderson. G. "Unfinished Voyages. Western Australian Shipwrecks 1622-1850". (University of WA Press, 1980).

Lemon. A & Morgan. M. " Poor Souls, They Perished". (Hargreen Publishing, 1986) (*Cataraqui*)

Loney.J. "ADMELLA".

Loney. J. "Australian Shipwrecks. Vols. 2-5".

Loney. J. "The Loch Ard Disaster".

Loney. J. "Maritime Australia."

Loney. J. "The Mahogany Ship" (Author, 1974).

McHugh. E. "Shipwrecks. Australia's Greatest Maritime Disasters". (Viking, 2003).

Mudie. I. "Wreck of the ADMELLA". (Rigby, 1966)

Nash. M. "Cargo for the Colony" (Navarine Publishing, 2001) (*Sydney Cove*)

Page. M. "Fitted for the Voyage".

Parsons. R. "Southern Passages".

"South Australian Shipwrecks".

"Ketches of South Australia".

Playford. P. "The Wreck of the Zuytdorp" (WA Historical Society, 1959).

Sexton. R. "Before the Wind. The Saga of the Star of Greece".

Society for Underwater Historical Research publications:

"Wardang Island. Graveyard of Ships".

"LOCH VENNACHAR Expedition Report".

"The Wreck of the LADY KINNAIRD".

Sprengel.W. "The Ecstasy and Agony of Guichen Bay". (*Koenig Wilhelm 11*)

Stone. P. "Encyclopedia of Australian Shipwrecks and Other Maritime Incidents." (Oceans Enterprises, 2006)

Web sites

These are too many web sites to list, and in most cases use of a search engine will quickly bring up many relevant listings. The National, and State, Libraries and museums web sites are treasure troves of information. The following sites, however, are particularly useful.

www.en.wikipedia.com (General research)

www.findingsydney.com (HMAS *Sydney*)

http://oceans1.customer.netspace.net.au/easw.html (Encyclopedia of Australian Shipwrecks)

http://www.environment.gov.au/cgi-bin/heritage/nsd/nsd_list.pl?state=SA (National Shipwrecks Data Base)

http://www.environment.gov.au/heritage/shipwrecks/australian.html

http://www.museum.wa.gov.au/collections/maritime/march/shipwrecks/Wreckfinder/menu1.html (WA Maritime Museum – old Dutch wrecks))

http://www.abc.net.au/backyard/shipwrecks (ABC Radio shipwreck series)

http://nla.gov.au/ (National Library of Australia)

http://www.awm.gov.au/ (Australian War Memorial)

http://www.flagstaffhill.com/ (Victorian shipwrecks)

Acknowledgments

Many readers of my earlier books on South Australian wrecks have sought more information about ships wrecked around the whole coast of Australia. I am pleased that such interest has been aroused in this intriguing subject and have written this book to try and further document a little more of our maritime history.

Some original research has been carried out while writing this book, but in the main it has been compiled from secondary sources.

I acknowledge the reference extensively made of publications produced by the Society for Underwater Historical Research, Graeme Henderson, Jeremy Green, Ron Parsons, Rae Sexton, Chris Halls, Jack Loney, Peter Stone, Evan McHugh and Don Charlwood. The work of many others is identified in the extensive bibliography.

In addition the collections in the State Libraries of South Australia, New South Wales and Victoria particularly provided a wealth of information and photographs, as did the National Library and Australian War Memorial.

The Western Australian Maritime Museum personnel and publications have been an invaluable resource, particularly in respect to the early Dutch wrecks. Their fostering and ongoing support for the Australian Institute of Maritime Archaeology has also been vital, including through the online national shipwreck data base maintained by the Australian Government's Department of Environment, Water, Heritage and Arts.

Assistance with photography of relics from SA Maritime Museum and Flagstaff Hill Maritime Village at Warrnambool has also been appreciated.

Many authors of maritime books have undertaken the hard slog of earlier research.Their efforts are appreciated and anyone interested in further reading is encouraged to refer to the books listed in the bibliography.

ABC Radio through its broadcasts and website has been the source of information to me, and has played a valuable role for the community through its Shipwrecks series.

In addition to photographic collections accessed, the assistance of Linda Cowan, Ken and Debbie Smith, Steve Crouch, Helen Pavlou and Peter Roberts has been appreciated as they have assisted with recent pictures from NSW, WA and Victoria.

Much of the material relating to South Australian wrecks comes from two of my earlier (now out of print) books, "Shores of Tragedy" and "to save the lives of strangers".

Also thanks are due to my family. As an author who undertakes research and writing as a hobby their encouragement, assistance and support is wonderful.

Peter Christopher.

SHIP INDEX

PHOTO REFERENCES

Reference No.	Source	Page
map-rm00689a-m	NLA	7
B30033	SLSA	8
map t00570-m	NLA	9
map rm 3221	NLA	9
Investigator anchor	Adam Lee	9
map rm00750-m	NLA	10
dig 005170	NLA	13
dig 005164	NLA	14
Batavia relics	Debbie Smith	15,16,17
Tryall cannon	Debbie Smith	19
Zeewijk cannon	WAMM	20
Zeewijk relics	Debbie Smith	21
Coins	Adam Lee	22
Gilt Dragon relics	Debbie Smith	23
Elephant tusk	Adam Lee	23
a1528496		
Dunbar painting	SLV	26
H90.111/280	SLV	26
H90.111/280	SLV	26
Tombstones	Linda Cowan	29
001/001151	Gosford Library	31
001/001149	Gosford Library	32
001/001141	Gosford Library	33
001/001146	Gosford Library	33
003/003209	Brisbane Water Historical Society	33
a09721	SLV	34
133923	SLQ	35
IMP27/08/66/320	SLV	37
Loch Ard Gorge	Helen Pavlou	38
H96.160/89	SLV	42
H96.160/100	SLV	42
Loch Ard	SLV	42
Chalice	Helen Pavlou	48
H96.160/100	SLV	49
H91.108/2754	SLV	54
		58
AUTAS001124064270	SLT	59
A/S29/12/87/197	SLV	59
Diver and relics	Parks & Wildlife Services Tasmania	60
MAR RM 676	NLA	61
C4078:N1069	NAA	62
B41900	SLSA	67

Reference No.	Source	Page
Notification document	Adam Lee	68
Medal	Adam Lee	68
Cabin fitting	Adam Lee	68
Certificate of Award	Adam Lee	69
Admella plate	Adam Lee	69
PRG280/1/43/84	SLSA	71
PRG280/1/43/78	SLSA	72
Memorial	Robyn Ashworth	72
Compass holder	Adam Lee	73
Clan Ranald	Chris Halls	73
Star of Greece figurehead	Adam Lee	74
B17690	SLSA	74
c.1888, Adelaide	Art Gallery of S.A.	74
Documents of wreck	Adam Lee	78
PRG1373/16/70	SLSA	76
Underwater	SUHR	77
B17480	SLSA	80
PRG280/1/1/134	SLSA	82
Postcard	SUHR	83
Crew	John Boon	84
RPG280/1/1/129	SLSA	85
66925	SLQ	88
dig 005144	WA Museum	90
dig 006142	QLD Museum	90
AUSTS001126072149	SLT	92
H99.220/649	SLV	93
H29557	SLV	94
b50021	SLV	95
B46876	SLSA	97
A/S20/03/75/205	SLV	97
HMAS *Sydney II*	Finding Sydney Foundation	103
305987	AWM	104
Kormoran's gun	Finding Sydney Foundation	104
Sydney life boat	Finding Sydney Foundation	104
emblem	Finding Sydney Foundation	105
wreath ceremony	Finding Sydney Foundation	105
Sydney's crew	Finding Sydney Foundation	105
301162	AWM	106
Perth relic	David Burchell	108
AN013558	SLV	110
P00444_189	AWM	110
060096	AWM	110

Reference No.	Source	Page	Reference No.	Source	Page
126519	AWM	111	134960	AWM	115
012580	AWM	111	P05303_004	AWM	115
305088	AWM	111	042895	AWM	115
061952	AWM	113	AN21511971	NLA	118
Sextant in box	Adam Lee	120	map-f00503-m	NLA	121
043235	AWM	133	Author	Nicholas Cundell	132

Photos not listed above have been taken by the author.

GLOSSARY

ANMM – Australian National Maritime Museum

AWM – Australian War Memorial

FHMV – Flagstaff Hill Maritime Village

NAA – National Archives of Australia

NLA – National Library of Australia

SAMM – South Australian Maritime Museum

SLQ – State Library of Queensland

SLSA – State Library South Australia

SLV– State Library Victoria

SLWA – State Library Western Australia

SUHR– Society for Underwater Historical Research

SLT – State Library of Tasmania

WAMM– Western Australian Maritime Museum

Dunbar wrecksite New South Wales

About the author

Peter Christopher has had a keen interest in maritime history, and has been diving on shipwrecks, for over 40 years. During this time he has been President of both the Underwater Explorers Club of S.A. and the Society for Underwater Historical Research (SUHR), and a Council member of the Australian Institute for Maritime Archaeology.

In the early seventies the author was a member of the South Australian Government's Inquiry into Sinkhole Deaths.

Peter has dived on wrecks in South Australia, Victoria, Queensland and Western Australia, the latter in conjunction with the W.A. Maritime Museum.

In 1982 he was actively involved with the Department of Environment and Planning and S.U.H.R. in the staging of the Second Southern Hemisphere Conference on Maritime Archaeology.

National recognition of Peter's contribution to the field of maritime archaeology was made in the form of an Award presented to him by the Commonwealth Minister of Home Affairs and Environment during this Conference.

Peter Christopher is the author of six other books, including the popular "Australian Riverboats. A Pictorial History" first published by AXIOM in 2006.

He is still active in research and diving activities, and is constantly upgrading an extensive computer data base of shipwreck, riverboat and maritime information which he has established.

As an author with a full time job, as well as undertaking research and writing, Peter still finds time to mess around in boats.

Peter Christopher lives in Adelaide.